BURIED PIPELINES FOR SURFACE IRRIGATION

BURIED PIPELINES FOR SURFACE IRRIGATION

Robert van Bentum
and Ian K. Smout

Intermediate Technology Publications
in association with
The Water, Engineering and Development Centre
1994

IT Publications, 103/105 Southampton Row, London WC1B 4HH, UK.
WEDC, Loughborough University of Technology, Leicestershire LE11 3TU, UK.

A CIP record for this book is available from the British Library.

ISBN 1 85339 187 5

A photographic supplement to this publication entitled 'Photographs of Buried Pipelines for Surface Irrigation' is available from WEDC on request.

Typeset by Karen Betts
Designed by Rod Shaw

Printed by SRP Exeter

CONTENTS

ACKNOWLEDGEMENTS

This publication originates from funding support by the Engineering Division of the Overseas Development Administration, (ODA) for research into buried pipe distribution systems for surface irrigation (contract R4575).

The authors wish to acknowledge the data and assistance received from many projects and individuals involved in the design and construction of buried pipe distribution systems. Particularly valuable assistance was received from Mott MacDonald International, as consultants to the Bangladesh Agricultural Development Corporation, on the IDA DTW II Project (Bangladesh) and through their UK office, as well as from Tangail Agricultural Development Project (GTZ funded) in Bangladesh.

Valuable data and assistance was received during the research work and field visits to existing buried pipe systems from a wide range of organizations and individuals including:

Bangladesh Agricultural Development Corporation.
Barind Integrated Agricultural Development Project, Rajshahi, Bangladesh.
Howard Humphreys and Partners, Thailand.
Indo-Dutch Tubewell Project, Monitoring and Appraisal Cell, Uttar Pradesh, (India).
Jain Industries in Jalgaon, Sendhwa, Indore and Lucknow, (India).
Ribloc Australia Limited, Thailand office.
Royal Irrigation Department of Thailand, Bangkok and Sawankhalok.
Rural Development Academy, Bogra, Bangladesh.
Shahada Sugar Co-operative, Maharashtra, India.
USAID personnel in Delhi and Madhya Pradesh, India.
World Bank Uttar Pradesh Tubewells Project II, Lucknow, India.

Thanks are also extended to all the individuals and organizations too numerous to mention here who provided copies of reports, information and personal comments which have provided invaluable background material. In particular the authors and publishers wish to thank the following who have kindly granted permission for the use of copyright material.

American Society of Civil Engineers for selected figures from the proceedings of the ASCE. Symposium on planning, operation, rehabilitation and automation of Irrigation water delivery systems. ASCE Irrigation and Drainage Specialist Conference 1987.
Food and Agriculture Organisation of the United Nations for selected tabulated data figures from their Irrigation and Drainage Paper No. 44. Design and optimization of irrigation distribution networks and the FAO Investment Centre Technical Paper 3/2 entitled Design and operation of irrigation systems for smallholder agriculture in south Asia. 1986.
Hydraulics Research UK. for Figures 4.2, 4.3 and 4.4. from Charts for the hydraulic design of channels and pipes. Fifth Edition. 1990.
Mott MacDonald International Ltd for a range of figures from their Deep Tubewell II Project Main report and Trainer's Guide. 1992.
Soil Conservation Service of the United States Department of Agriculture for numerous diagrams of buried pipe system structures from their National Engineering Handbook Section 2, Part I. 1967.
John Wiley and Sons Inc. for Figure 5.22 from 'Principles of farm irrigation design'. L.G. James. 1988.

We have attempted to trace all the copyright holders. We apologise if any have been inadvertently overlooked.

At a personal level, we would particularly like to thank Linda Brown, David Donald, Toby Harrison (posthumously), Guy Jones, John Merriam, Karim Mridha and Harun-ur-Rachid for their various contributions to the research, and Karen Betts, Neal Burton and Rod Shaw for producing the book.

PREFACE

IRRIGATION is a key element in crop production in all continents, and surface irrigation using canals accounts for some 90% of the world irrigated area. It has been practised in various African and Asian countries for hundreds or thousands of years, and systems currently in use were developed by governments of all hues, by agrobusiness, by groups of farmers and by individual farmers. Such systems vary in size from less than 1 hectare (ha) to more than 1 million ha, from diverse sources (river, lake, reservoir or well).

With recent urbanization, industrialization and agricultural development, water for irrigation is scarcer and farmers' crops are more dependent on optimum irrigation. Under these conditions two shortcomings of surface irrigation systems become more important: firstly, a high proportion of water is lost between the source and the crop, and secondly, deliveries of water to individual farmers are not optimum in quantity or timing or reliability. The extent of these shortcomings depends on the physical infrastructure and management of the system, and its particular environment, but the problems are found throughout the world.

Irrigation engineers are aware of various options available to tackle these problems and achieve better performance from surface irrigation systems, including lined canals to reduce losses, control structures and systematic management to improve deliveries, and "demand"-type systems to enable farmers to control their own water supplies. A different approach is to adopt sprinkler or trickle irrigation systems, supplied through high-pressure or medium-pressure pipelines. These can provide more flexible deliveries with lower losses, but they tend to have considerably higher capital and operation and maintenance costs in the South than surface irrigation systems, requiring high-value (usually export) crops.

The aim of this book is to present a less well known alternative technology, the low-pressure buried pipeline system for surface irrigation. This offers improvements over conventional surface irrigation systems with canals, without the sophistication and high cost of pressurized sprinkler and trickle irrigation. It is thus an intermediate technology which may be appropriate on many irrigation schemes in the South. It will normally be more efficient and more flexible in operation than conventional canal systems, with fewer environmental problems. The costs are considerably lower than for sprinkler or trickle irrigation systems, and comparable to those of lined canals.

We hope that the book will encourage irrigation planners and practitioners to consider using buried pipeline systems, and that it will provide the information they need to follow this through. One of the main constraints on adopting this technology has been that experience and information on it was limited to a few countries, and we hope that this book will partly overcome that. The references in the text should help the reader to understand the origin of the material presented, and to follow it up if desired.

The book is based on research into buried pipe distribution systems for surface irrigation carried out by the Water Engineering and Development Centre (WEDC) of Loughborough University of Technology in collaboration with the Bangladesh Agricultural Research Institute (BARI). The work was undertaken from December 1989 to August 1991 and was mostly funded by the Overseas Development Administration of the UK (ODA).

The research project examined buried pipe networks for distributing water at low pressures for surface irrigation. It included a desk study and review of literature, a field visit programme to pipe systems in India, Bangladesh and Thailand, and fieldwork undertaken by our collaborators on eight buried pipe systems from deep tube-wells in Tangail District, Bangladesh. The desk study sought to compare buried pipe systems with more conventional unlined and lined open channel systems. Both published and unpublished material was examined, with a particular focus on documentation of field experience of local irrigation practitioners with the design, operation and construction of pipe systems.

Following on from contacts made during the literature search and review, field visits were undertaken to buried pipe distribution systems, utilizing important innovations in design and operation, in India, Bangladesh and Thailand. These have been described by van Bentum (1992).

The Bangladesh field research work aimed to quantify specific benefits such as reduced transit losses, reduced land take for the distribution system and real advantages for operating and maintaining the system. Results from this work have been reported by Rashid et al (1992) and Mridha (1993).

This book is based on the research findings and is intended to provide practitioners with guidelines on the selection, design and construction of buried pipe distribution systems for surface irrigation, particularly in the South. These guidelines apply to low-pressure pipe systems, i.e. those with operating pressures not greater than 10m (i.e. 1 bar or 1kg/cm^2) and usually less than 6m (0.6 bar). Reference is also made to some medium pressure systems which supply surface irrigation schemes. We have drawn extensively on the design and construction experience gained by staff on projects in India and Bangladesh, but the guidelines apply to the full range of pipe systems with varying water sources, pipe materials and pipe layouts.

The book aims to help with identifying the situations in which pipe systems are attractive alternatives to canal systems. The range of pipe systems and their component structures are described and classified to enable selection of the most appropriate pipe system under a range of criteria. This is assisted by the use of a number of flow charts.

Specific recommendations on design and construction are made where they are of relevance to buried pipe systems or where they deviate from accepted practice for canal systems. Emphasis is given to the design of the pipe layout, particular aspects of hydraulic design and design considerations for component structures.

The first draft of this book was as a report to ODA entitled "Guidelines on Design and Construction of Buried Pipe Distribution Systems for Surface Irrigation" (1991). This has been substantially rewritten for this book.

A word of caution. Although we want to encourage wider consideration of the promising technology presented in this book, we do not advocate technological fixes and top-down development. Rather we believe that successful irrigation development requires farmer participation at all stages from planning, through design and construction, to operation. The workings of pipe systems are unfamiliar to farmers (even more than to many irrigation engineers) and particular efforts will be required to achieve effective farmer participation. We have tried to provide some pointers to this in the text, but it will be for practitioners to work out suitable procedures in conjunction with farmers.

We wish you luck, and we would welcome information on your experiences, as well as comments, corrections and additions to the material presented in this book.

Ian Smout
Water, Engineering and Development Centre,
Loughborough University of Technology, 1994

CHAPTER 1

BURIED PIPE DISTRIBUTION SYSTEMS FOR SURFACE IRRIGATION

Introduction

In this chapter standard terminology on irrigation is introduced, and the various options for irrigation distribution are described. This is followed by a description of typical buried pipe distribution systems, and a worldwide review of the adoption of buried pipe systems for irrigation. A comparison is presented of the advantages of open channel and buried pipe systems. Pipe materials and methods of manufacture are summarized. The chapter ends with a detailed classification of buried pipe distribution systems.

Distribution systems for surface irrigation

The area which can be irrigated from a canal system (usually by gravity) is termed its command area. This is defined by Bos (1985) in terms of area units of different size, with each a subset of the other. In ascending order these are quaternary or block, tertiary, secondary or lateral, and finally scheme or project area. A schematic presentation of the more general terminology is detailed in Figure 1.1.

The quaternary unit can be defined as the area that can be irrigated efficiently by one farmer using a continuous flow through an outlet from the tertiary canal or pipeline. In many cases the flow will be used consecutively by several farmers. The tertiary unit is a group of two or more quaternary units and receives water from the secondary canal or pipeline, which is supplied from the main canal.

The movement of water through the units of the irrigation system from the source to the crop is commonly considered as three separate operations: conveyance, distribution and field application (Bos and Ngutren, 1990):

- *conveyance* is the movement of water from its source through the main and secondary canals or pipe conduits to the tertiary offtakes;
- *distribution* is the movement of water through the tertiary and quaternary canals or pipe conduits to the field inlet;
- *field application* is the movement of water from the field inlet to the crop.

The classification was developed for large irrigation systems, and not all of these levels necessarily exist in a particular scheme. Small schemes may consist of a distribution system supplied directly from the source, with no conveyance system.

The tertiary distribution system begins at the point of water entry to the tertiary unit, which is typically 40 to 100 ha in area. The flow to be delivered to each quaternary unit (the delivery flow) will be determined by the particular flow rate (Q_f) which the farmer can handle efficiently with the irrigation system he or she is using — for example about 30 l/s for a quaternary canal delivering water for most surface irrigation methods, or 2-8 l/s for

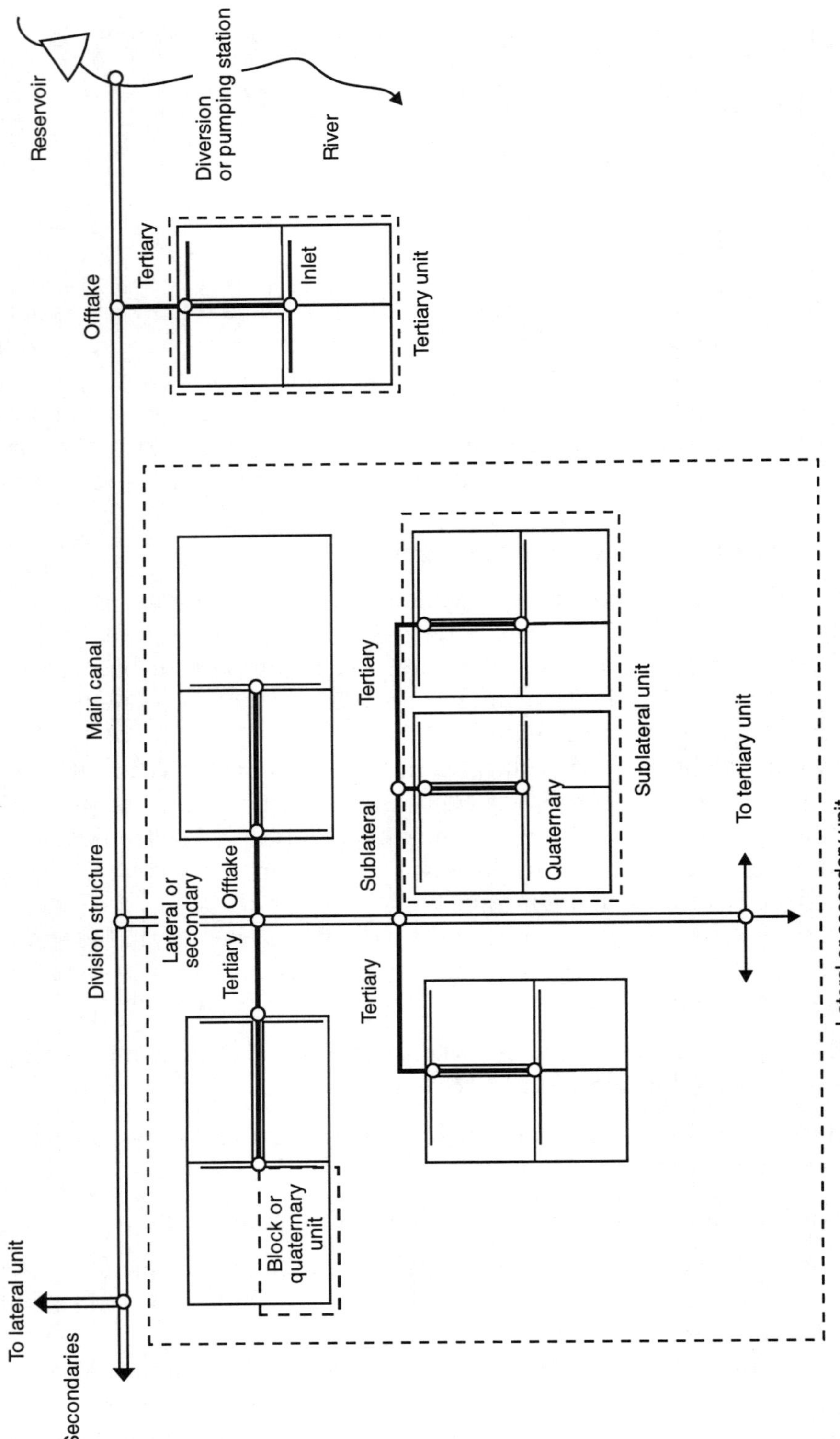

Figure 1.1. Schematic presentations of irrigation system definitions

Source: Bos (1985)

a hose attached to a hydrant. The inflow (Q_i) to a tertiary distribution system will either be equivalent to the desired delivery flow (plus losses), or a multiple of this, necessitating division of flow within the system or at the outlet.

The tertiary distribution system may comprise open channels (earth or lined) or pipes or a combination of these.

Earth channels are the traditional method of distributing irrigation water, and are used on both indigenous and modern systems. Earth channels may have high water-loss rates, but those which are built from well compacted clayey soil can have low seepage losses as long as they are well maintained.

Lining has been the conventional strategy for reducing losses. There is a wide range of lining alternatives, ranging from low permeability clay or earth layers, through plastic membranes and asbestos sheets to hard surface linings such as concrete or brick (Deacon, 1984).

This book is concerned with the use of low-pressure buried pipelines for distribution systems on surface irrigation schemes. Such systems are quite widespread, particularly in USA, China and India, and serve about 4% of the world irrigation area. Low-pressure pipelines are also occasionally used for conveyance instead of primary and secondary canals (Galand, 1989, Merriam, 1987a and the American Concrete Institute 1990).

Description of typical buried pipe distribution systems

Overview

A typical buried pipe distribution system is a tertiary level system which receives water from a canal, reservoir or pump and distributes it over a command area of 10 to 100 ha via a loop or branch layout of buried pipes to a number of outlets. Each outlet supplies water to one or more quaternary distribution systems, which comprises either earth channels or surface pipes or hoses. The only structures which are visible above ground are inlet structures which direct the water into the pipeline at the head of the system, air vents, stand-pipes (for valves and surge risers) and outlets.

Schematic illustrations of typical branch and loop buried pipe distribution systems are shown in Figures 1.2 and 1.3 (MMI, 1992). Definitions of standard structures, based on their function, are given in Chapter 2.

Most low-pressure pipe systems have until recently been constructed from non-reinforced concrete pipe but plastic (uPVC) pipe materials are now increasingly being used. In some rare situations, for reasons of cost and availability, asbestos cement pipe is sometimes used. Pipe diameters range from as small as 150mm for uPVC pipe up to 600mm for the larger concrete pipes. Low-pressure pipe systems with larger diameter pipes also exist but these tend to be secondary conveyance pipelines rather than tertiary distributary systems.

Most buried pipe systems operate with a maximum pipe pressure which does not exceed 5m and is frequently less than 3m hydraulic head. The non-reinforced or nominally reinforced concrete pipe widely used for these pipelines cannot withstand pipe pressures (inclusive of surge pressures) which exceed 5m. However, operating pressures up to and in excess of 10m are acceptable for pipe systems using uPVC pipe. The higher pressures and velocities allowable with uPVC pipe enable a reduction in pipe diameter to be made for the same flow rate, thus helping to reduce the cost of the pipeline.

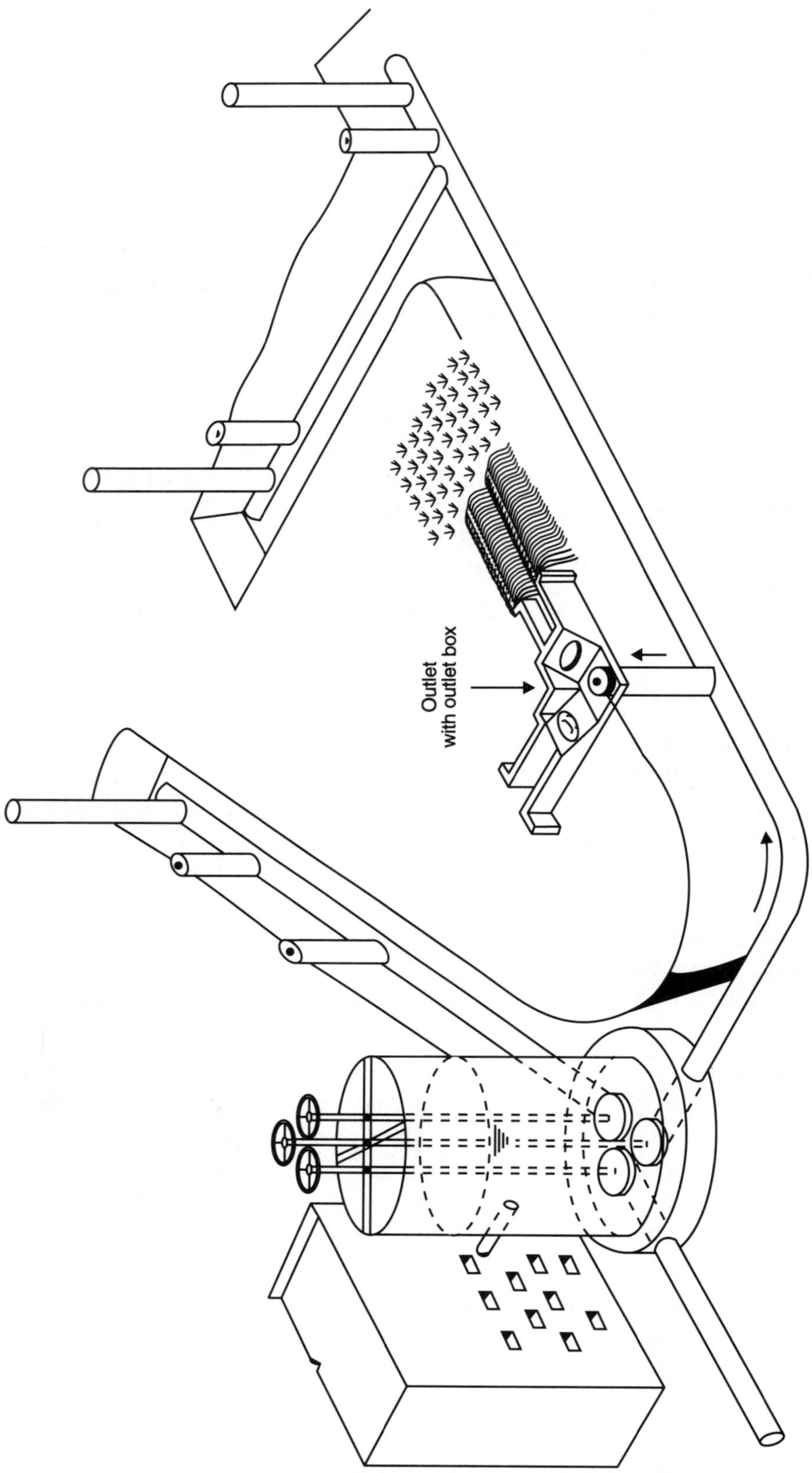

Figure 1.2. Schematic illustration of a branched buried pipe irrigation system

Source: MMI (1992)

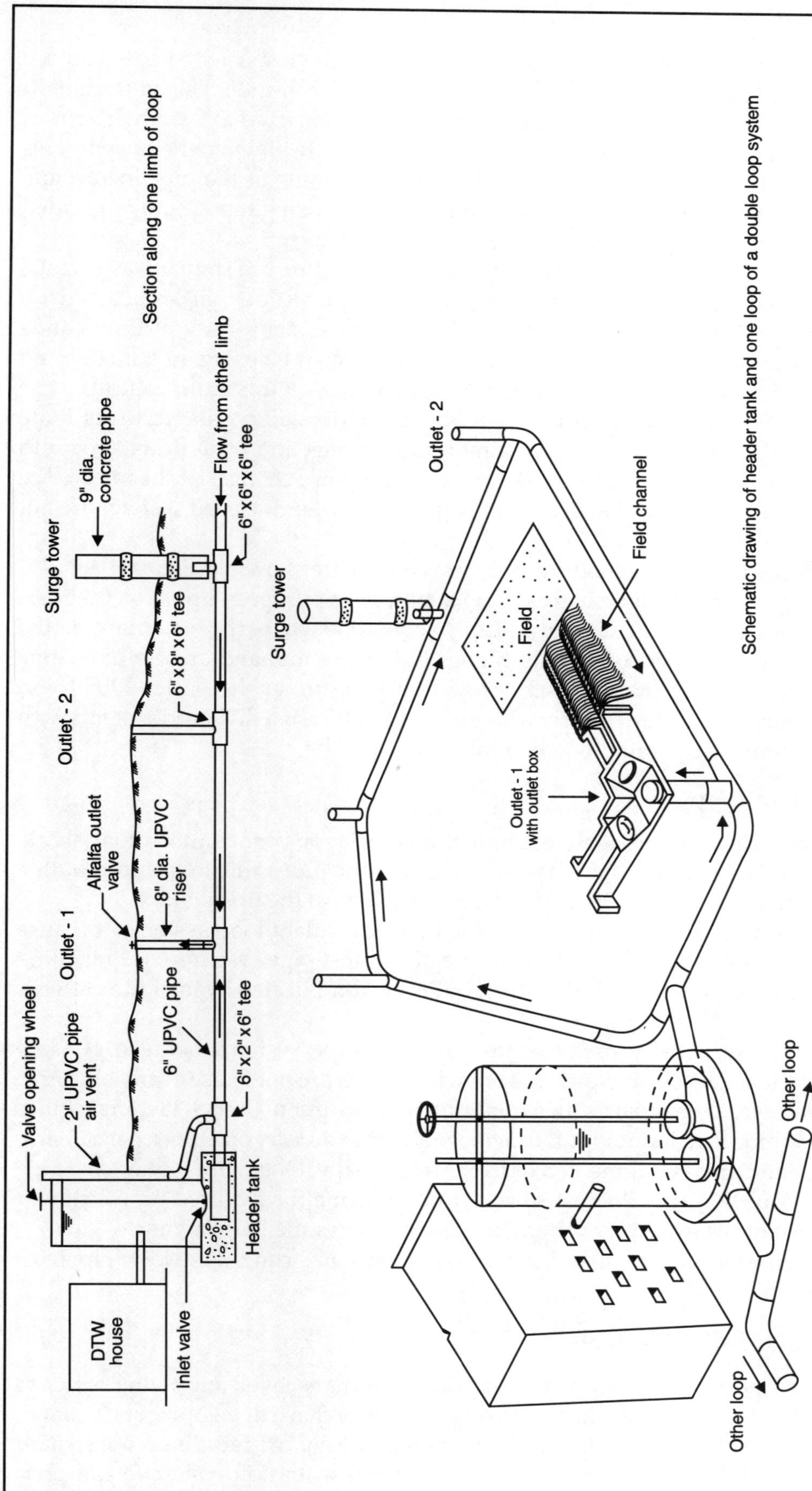

Figure 1.3. Schematic illustration of a double loop buried pipe irrigation system

Source: MMI (1992)

Pressure control

Buried pipe systems are defined as closed, open and semi-closed with regard to the method of pressure control. Closed pipe systems are the most common. The typical closed pipe system is one where all the outlets are hydraulically connected to the water source, so the opening and closing of an outlet affects the pressure at the inlet to the system. This allows the system to be operated on demand, within the limits of the pipeline design. Closed pipe systems can be used on near-flat, undulating or rising topography provided the pressure available at the inlet to the pipe system is sufficient.

Open and semi-closed systems are used in falling topography, in a similar way to canal systems. In an open pipe system, open stand-pipes with overflow baffles act as drop structures to dissipate the excess hydraulic head within the system. The overflow stands serve to disconnect sections of the pipeline so that opening and closing an outlet has no effect upstream of the drop structure (as in a canal system with upstream control).

In the semi-closed system, float valves are used instead of overflow stands to dissipate the excess hydraulic head. By providing a hydraulic connection from downstream to upstream, they allow changes in flow to be sensed as pressure changes at the inlet to the pipe system (as in a canal system with downstream control). Semi-closed systems are not widely used.

Examples of the different methods of pressure control are shown in Figure 1.4.

Pipe systems are also distinguished as gravity supply or pumped supply on the basis of the origin of the driving head. Gravity supply is possible where the water level at the source (canal or reservoir) is sufficiently higher than the command area to overcome friction losses and provide the necessary operating pressure at the outlet. Otherwise pumped supply is required, particularly where groundwater is used. Closed systems with pumped supply are widely found on deep tube-wells in Asia.

Layout: branch or loop

Buried pipe systems may comprise either a branch or loop layout, or a combination of both. In branch layouts the full outlet discharge is sent down one pipe route to the open outlet, while in a loop, the discharge is divided between two legs to the open outlet.

Branch systems require larger diameter pipe than equivalent loop systems, because they must take the entire outlet discharge in one pipe. Most pipe systems use branching pipe systems, particularly where the layout of outlets is complicated by intricate patterns of land ownership.

Branch layouts are usually preferred for concrete pipe systems because pipeline leakage can be reduced by closing pipeline branches which are not in use at any one time. This also facilitates leakage repair with a minimum of disruption. Branch layouts require considerably less pipe than loop layouts to achieve a similar density of outlets, particularly where the command area is fragmented or irregular in shape. Only with higher-cost pipe materials, such as uPVC, does the cost-saving from reducing the required pipe diameter outweigh the cost of the extra pipe length necessary to form the loop layout.

Only closed pipe systems or closed parts of other systems can utilize loop pipelines within their network.

Quaternary distribution systems

In some cases buried pipe systems extend to the quaternary level, supplying water to individual landholdings and individual furrows in some orchard developments (Galand, 1989); similarly the outlets from the pipe system may discharge into lined quaternary channels. However it is more usual for the quaternary distribution system to comprise earth channels, or movable surface pipes or hoses.

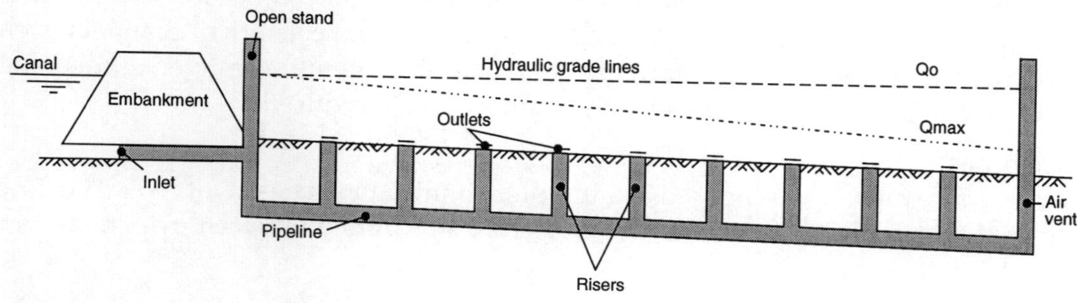

(a) Closed system

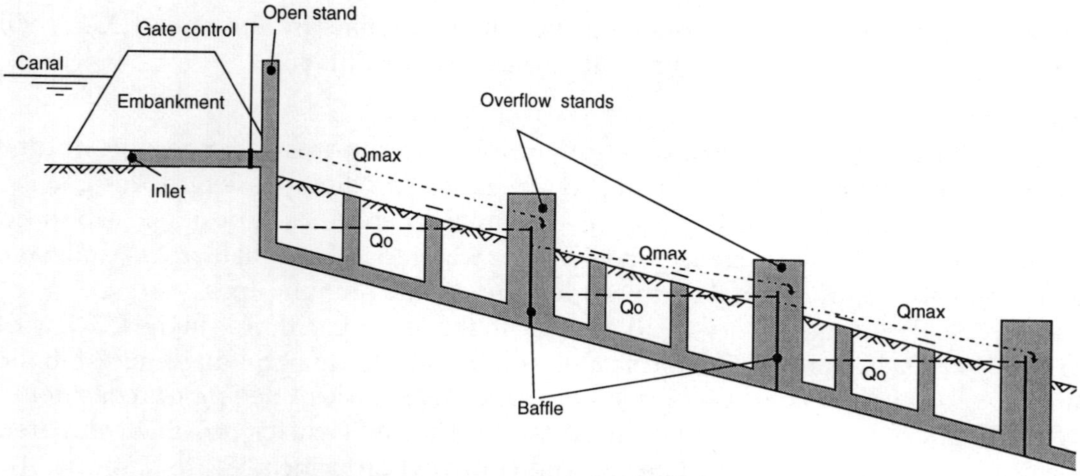

(b) Open systems with overflow stands

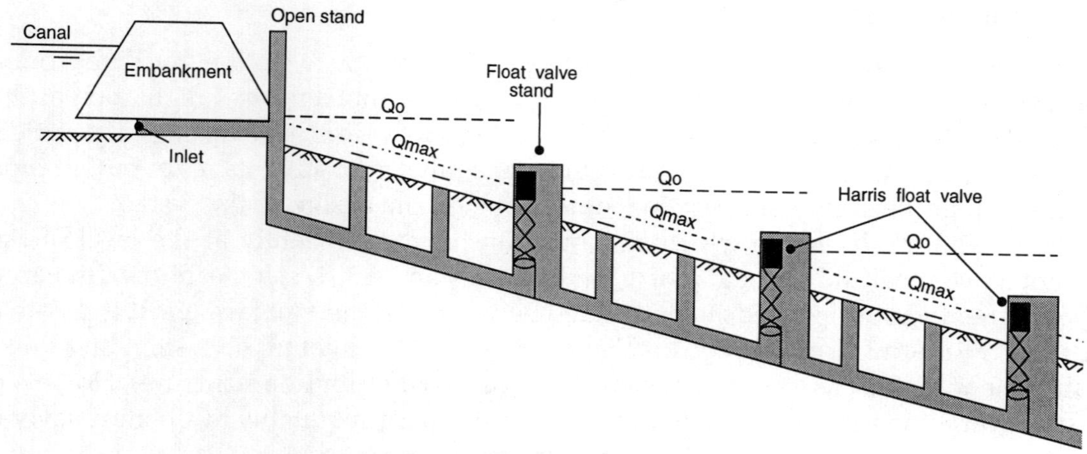

(c) Semi-closed system with float valves

Figure 1.4. Buried pipe systems under gravity supply

Source: Baudequin et al (1990)

Earth channels
Significant losses of water can occur in quaternary level earth channels. They tend to be constructed with less care than their tertiary level equivalents, a reflection of both their often temporary nature (constructed on an annual basis) and the length of channel which each irrigator usually has to construct and maintain. Where attention is given to adequate compaction and shape, seepage losses can be substantially reduced.

Surface pipes
Movable surface pipes are composed of aluminium or ultraviolet-resistant PVC. The flow is delivered to one plot or field, and the pipes are only shifted between irrigations, not while water is flowing.

Surface hoses
Flexible surface hoses will generally be of smaller diameter than surface pipes, allowing water to be redirected to different field plots without interrupting the flow of water. Flow rates are generally low (2-10 l/s) and such hoses are mostly used to irrigate horticultural crops. Reference is made to the use of hoses on buried pipe systems in both China (DST, 1990) and the Kallada Irrigation Scheme in south India (Campbell, 1986).

Gated pipes
Gated pipe, typically made from aluminium or uPVC, comprises a closed pipe or tube with controllable orifices (gates) located at regular intervals along its length, designed to coincide with a particular furrow spacing. Gated pipes are specifically designed to be used at the outlets of buried pipe systems for delivering water to individual furrows, and have tended to replace the traditional earth header ditches and siphon pipes.

Galand (1989) reports the very rapid increase in the use of gated pipe in the USA, with more than 360 000 hectares in California alone, and indicates the growing interest in the technique among irrigators in southern France. In both countries gated pipe is considered attractive because of the lower labour input and higher efficiencies possible compared with earth header ditches. Typical design and construction principles are described by Watts and Smith (1985), Eisenhauer (1990) and Kruse et al (1980).

Operation of buried pipe distribution systems

The inflow of water at the inlet of a buried pipe distribution system may come from a reservoir, canal, pump or pipeline. Operation of the system comprises regulation of the inflow and setting of the outlets, so as to give satisfactory discharge through the outlets without water spilling from the system. One of the important advantages of buried pipe systems is their flexibility of operation, depending on the design of the system.

The maximum discharge which a farmer can handle efficiently at the level of the quaternary channel and the field plot (Q_f) is generally about 30 l/s. If the inflow discharge (Q_i) is higher than this, it needs to be divided between a number of farmers. It is possible to divide the inflow discharge accurately using weirs at the inlet to the system (in a tank) or after the water discharges from the outlet valve (at a distribution structure). However, pressure variations make it difficult to divide the discharge equitably by opening several outlets on the same pipeline and this is not recommended.

The options for operation of a buried pipe distribution system depend on its type: closed or semi-closed or open system, gravity or pumped supply with diesel- or electric-powered pumps.

Manual regulation of the inflow is possible on all closed and semi-closed pipe systems and is the only option on open pipe systems. On closed and semi-closed pipe systems control of one gate or valve at the inlet to the pipe system is all that is necessary. On open pipe systems however the gates and valves must be carefully set throughout the system to balance the inflow with the outlet discharge, and these must be adjusted repeatedly to match changes in outlet discharge, if spillage of water is to be avoided.

On closed and semi-closed systems with gravity supply, on-demand operation of outlets is possible provided sufficient water and head is available at the inlet, up to the limit of the system design. No water will flow unless an outlet is opened. The inlet is usually fitted with a gate or valve so that the inflow can be shut off manually or restricted to below the design capacity of the pipe system.

On closed and semi-closed systems with pumped supply, the method of regulating inflow to the pipe system depends largely on the type of pump drive. For pumps driven by diesel engines the inflow and the outlet discharge are matched by varying the motor speed and altering the number and setting of outlets which are operating. For pumps driven by electric motors, the inflow is matched to the outlet discharge by automatically switching the pump off and on, in response to changing water levels in a tower or tank at the inlet to the pipe system, or by manually altering the number and setting the outlets which are operating.

The extent and history of buried pipe systems in surface irrigation

Although precise statistics on the proportion of surface irrigation using buried pipe distribution systems do not exist, reasonable estimates can be made based on data concerning world irrigation and known areas of pipe systems in some of the major irrigating countries of the world.

Field (1990), based on data from a number of FAO and World Bank publications, reports the total area of world irrigation at around 256 million hectares, with 94% of this being surface irrigation. Estimated areas of the different types of irrigation for both developed and developing countries of the world are summarized below in Table 1.1.

Table 1.2 shows estimated areas of buried pipe distribution systems, compiled during this research. In respect of the area of low-pressure pipe systems, USA has clearly the largest area at 7.3 million hectares (Baudequin et al, 1990) followed by China (DST, 1990) and India. These area estimates are based on known areas of development with an allowance for some private investment outside government control, and are likely to underestimate the area of low-pressure pipe systems. Although buried pipe systems in their varied forms comprise nearly 43% of all surface irrigation in the USA (Baudequin et al 1990),

Table 1.1 **Summary of world irrigation areas ('000 ha)**

System	*Developing countries*		*Developed countries*		*Totals*	
Surface (excluding low-pressure pipes)	180 255	(97%)	46 628	(68.6%)	226 883	(89.5%)
Sprinkler	1 500	(0.85%)	12 592	(18.5%)	14 092	(5.5%)
Micro-irrigation	200	(0.15%)	1 000	(1.5%)	1 200	(0.5%)
Low-pressure pipes (estimated)	3 685	(2%)	7 740	(11.4%)	11 425	(4.5%)

Source: Field, 1990 and Table 1.2

Table 1.2 Estimates of individual country areas of buried pipe systems ('000 ha)

Developing countries		*Developed countries*	
China	2500	USA	7310
India	1000	France	200
Bangladesh	10	Japan (estimated)	60
Africa	100	Australia	40
Nepal	5	Spain, Portugal	130
South-east Asia	20		
South America	50		
Totals	*3685*		*7740*

Source: van Bentum, 1992

when taken in a world context they currently account for less than 5% of the total world irrigation area. Thus pipe systems could be considered an American specialization, though the situation is quickly changing.

While surface irrigation remains the dominant type of irrigation, the demand for greater water-use efficiency, driven by higher unit water costs and water-resource constraints, has led to a shift in emphasis particularly in developed countries towards sprinkler and micro-irrigation technologies. Pimley and Fischer (1990) ascribe this change to the poor performance efficiency of labour-intensive surface irrigation using inexpensive water.

Plusquellec et al (1988) also argue that one of the driving forces for sprinkler and trickle irrigation development has been the interest and investment on the part of private sector manufacturers of irrigation equipment. They state that "surface irrigation with little opportunity for marketable hardware has aroused only limited interest from the private sector."

The move from surface systems to sprinkler and micro-irrigation systems, although providing valuable savings in labour, requires corresponding improvements in yield and crop value to cover the associated higher capital and energy costs. These have been realized in higher-income countries, with their higher labour costs. However, for the majority of smallholders in developing countries growing staple crops with existing surface irrigation schemes and low labour costs, returns can seldom justify the capital expenditure associated with such a change. Their need remains for more reliable irrigation which can respond to their changing water needs.

Merriam (1987c) argues that buried pipe distribution systems represent an intermediate solution between lower-cost earthen channels and higher-cost sprinkler and micro-irrigation systems. Existing canal systems can be upgraded to more flexible operation by constructing intermediate reservoirs along the secondary canals, supplying buried pipe distribution systems.

Comparison of open channel and buried pipe distribution systems

One of the important reasons for interest in buried pipe systems is the evidence of poor performance, high cost and short life of open channel alternatives, which is described in the following sections. These shortcomings are particularly important where improved irrigation deliveries are required to raise levels of agricultural productivity on the current area of irrigated land.

However it must be remembered that social, organizational and management constraints on irrigated production levels may be as or more important than technical problems, and need to be considered in any successful intervention to improve agricultural performance. For example Figure 1.5 shows the factors which were found to limit the command area on tube-wells with buried pipe distribution systems in Bangladesh (Rashid et al, 1992).

Such constraints may be independent of the type of distribution system — for example marketing problems — but there may be important interactions in some situations. For example earthen canals may require greater co-operation between farmers in rotating deliveries systematically, and in maintenance, and they may be more prone to unauthorized abstractions ("water stealing") than lined canals or pipes. The issues require site-specific analysis in each case.

Performance of open channel distribution systems

Earthen channels

Unlined earth channels tend to have low distribution efficiencies, with water lost to seepage through the banks and bed, leakage through poor structures, animal burrows and root holes, and overtopping of banks. Important determinants of the magnitude of seepage losses appear to include soil texture, compaction during construction, maintenance and the level of the water table.

Other disadvantages of earth channels are the transit time required for water to flow to an outlet, and the susceptibility to weed growth which can reduce the capacity of the canal and further increase the transit time.

Low distribution efficiencies and high transit times of flow from the source to the field restrict farmers' flexibility of irrigation. Water losses and weed growth also bring risks to the environment from water-logging and water-related disease.

Lined channels

Lined channels have a greater capital cost than earth channels. Important reasons for lining canals are:

- reducing transit losses of water from the canal by seepage, leakage or overtopping;
- decreasing the transit time for water distribution;
- using steeper slopes, higher velocities and a smaller canal section;
- reducing land acquisition requirements (especially by using precast canalets);
- reducing weed growth and risks of diseases such as schistosomiasis;
- easing cleaning and maintaining the integrity of the canal section.

In practice transit losses from lined canals depend on the standards of construction and maintenance, and are higher than potential seepage rates suggested by laboratory measurements. Field work often shows a rapid deterioration in the performance of rigid linings. For example, Goldsmith and Makin (1989) describe field measurements on lined distributaries in the Punjab, where loss rates were shown to be similar to those of the earthen channels they replaced some years earlier. They distinguish between new channels and those more than four years old, of which "the poorest had loss rates higher than those measured in unlined channels." They attribute the major cause of these high transit losses to the development of cracking in the canal lining.

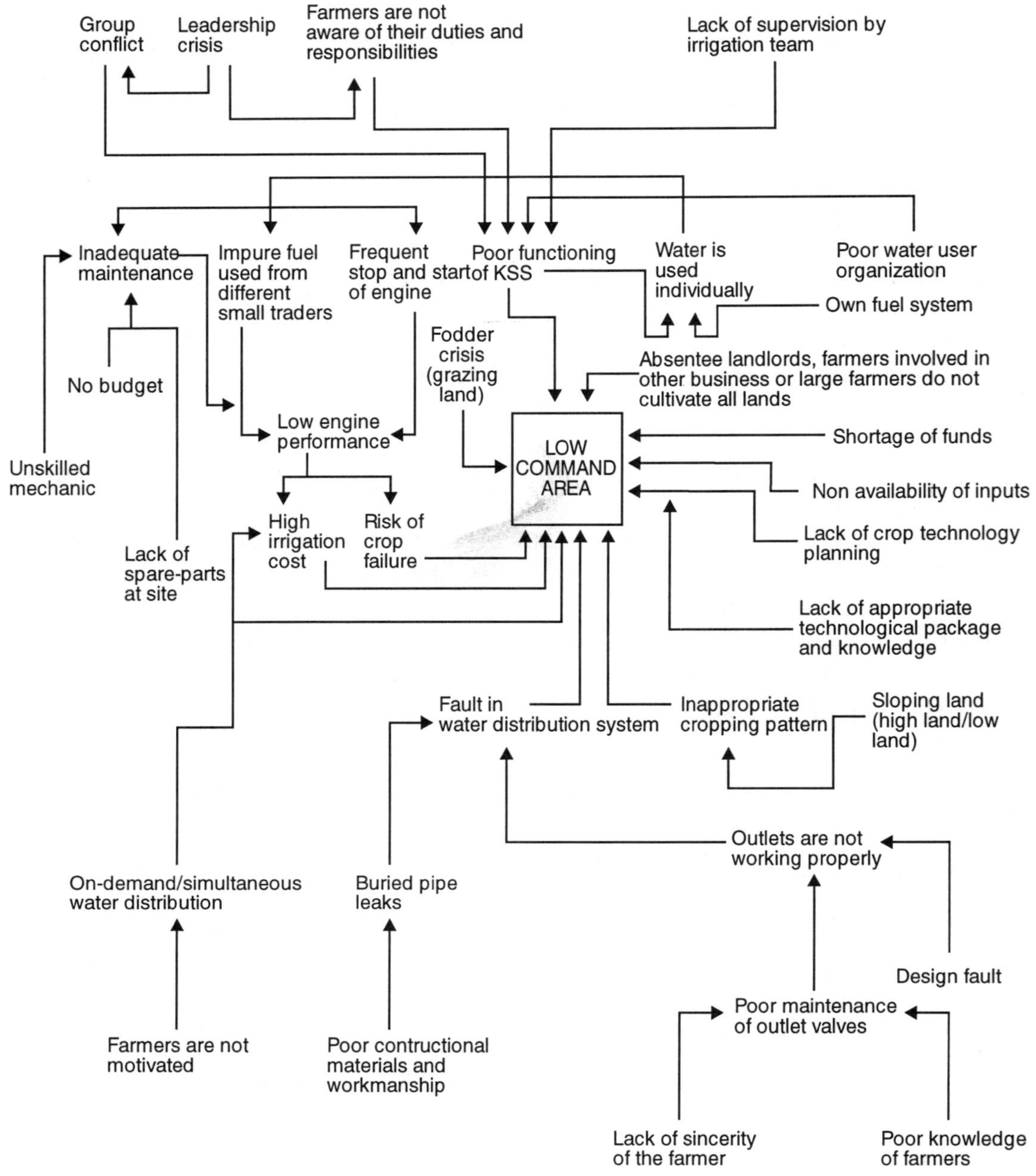

Figure 1.5. Flow chart of constraints to command area development

(Rashid et al, 1992)

The benefits and advantages of buried pipe systems

The benefits ascribed to buried pipe systems can be usefully divided into qualitative improvements in the delivery of water and operation of the irrigation system, and quantifiable savings in land, water and system costs.

Qualitative benefits

General

Important characteristics of buried pipe distribution systems are reduced water losses, greater flexibility and increased reliability. Reports from operating schemes demonstrate that these can help improve various components of agricultural production: irrigated area, crops grown, intensities and yields.

Campbell (1984) concludes that pipe systems in northern India assured flow delivery at the design discharge to the furthest irrigator with a minimum of losses and unauthorized diversions en route. Economic benefits were generated by the higher level of agricultural development (growing of high-value crops) which was made possible by the improved flow deliveries.

Gisselquist (1989) similarly reports successful crop diversification on buried pipe distribution systems in Bangladesh, benefitting from the systems' flexibility.

Other important qualitative benefits include:

- Short transit times enable the rapid shifting of irrigation water from one part of the command area to another during periods of fluctuating or low demand.
- Short transit times effectively reduce the distance between the farmer and the water source, decreasing the magnitude and importance of tail-end problems and water stealing.
- Larger delivery flows allow for quicker and more efficient field application.
- Reduction in the disruption to existing crops and any existing irrigation system during construction, resulting in fewer delays in obtaining agreement of the layout and alignment of the pipe system.
- Elimination of the need for expensive raised embankments and aqueducts to convey water across drainage features, gullies or depressions.
- Savings in water, which can be significant enough to enable the selection of a lower pumping duty, resulting in savings in both capital cost and operating costs.

Environment

Although little quantified information has been found on the environmental and health impact of buried pipe distribution systems, the documented savings of agricultural land, reduced water losses and elimination of suitable habitats for disease vectors will all have a positive impact, compared to open channels. The effects of buried pipe systems on some commonly reported environmental problems are summarized below:

- inefficient use of resources of land, water and energy: buried pipe distribution systems directly reduce the problems of loss of land resources and, through their lower transit losses, they also save resources (water and possibly energy used in pumping, and the resources used in construction);
- damage to land (particularly through waterlogging and salinity): reduced because less water is lost from a pipe system;

- damage to water resources (for example by drainage water degrading the quality of groundwater or surface water resources): quantities of drainage water are reduced because of the greater transit efficiency;
- displacement of previous land users: less of a problem because the land take is reduced;
- damage caused during construction: construction is quicker and less disruptive;
- spread of aquatic weeds and associated pests (e.g. rodents) and disease vectors: suitable habitats in the irrigation system are eliminated by pipe systems; drains are drier, less suitable habitats because of improved transit efficiencies;
- increase in human illness (particularly schistosomiasis, malaria): by elimination of vector habitats, pipe systems can assist with the control of these diseases.

Where irrigation systems are also used for domestic water supply, the reduced opportunity for contamination of the water supply once in the pipe system should result in improvement in water quality and therefore a reduction in the risk of waterborne ill-health. MMI (1990b) describes the impact on access to water for domestic purposes of converting an open channel irrigation distribution system into a pipe system. While good quality water becomes available at locations far from the water source, water is less accessible at the source if the discharge from the pump immediately enters the pipe system.

Disadvantages

The main reason for the non-adoption of buried pipe systems is the apparently higher capital cost, but a number of other disadvantages have been noted, including:

- the greater head required for a pipe system than for an open channel system;
- the fixed nature of the pipe system layout and outlet positions, which cannot be changed easily;
- unauthorized water use is more difficult to identify because the flow of water is hidden through most of the distribution system;
- the trenches dug during pipeline installation disrupt irrigation and cropping practices, often restricting system construction to short periods during the year;
- successful construction and maintenance of the pipe system requires skilled labour not always available in the village;
- the pipe system represents a new, more complex technology for the smallholder farmer to operate;
- unfamiliarity of the irrigation engineer and technician with design and construction of pipe systems.

Particular note should be taken of the need for capability in design, construction and operation, which are higher than the requirements for open channel systems. Institutional strengthening may be needed to facilitate the introduction of buried pipe systems.

Quantifiable benefits

The three main quantifiable benefits of buried pipe systems over their open channel alternatives are:

- reduction in water transit and distribution losses;
- reduction in the land area taken up by the distribution system;
- reductions in the maintenance and operating costs of the irrigation system.

Distribution and transit efficiencies

The overall distribution efficiency of pipe systems is considered as the product of the transit efficiency of the pipe system and the distribution efficiency of the quaternary channels. Higher transit efficiencies are obtained with pipe systems, but it may be difficult to improve the quaternary distribution efficiency.

The transit efficiency of the pipe system is the ratio of the water discharged from the outlet valve or valves to that which is entering the pipe system. Seepage losses can occur from the pipe body, the pipe joints, and the structures associated with the pipeline, for example outlets.

Leakage from the pipe body is not usually significant except where the pipe has been poorly manufactured or damaged. Adequate testing of the pipe prior to installation should ensure that faults are identified at an early stage. Losses from leaking joints are usually more serious. Special attention needs to be given to the installation of pipelines under roads to avoid high seepage losses (Rashid et al, 1992). It is also clear that leakage in low-pressure concrete pipe is greatest in the sections near the water source or inlet where operating pressures are highest.

Seepage loss from structures associated with the pipe system generally results from damage to above-ground structures such as header tanks and outlet valves which no longer seal tightly. It may also be caused by movement or settling of structures after construction. On the systems which have been studied very few structures have significant seepage losses after one season of operation, apart from outlet valves which have been broken. Total seepage loss from outlet valves will depend on the number of leaking valves in the system, the pipe layout (branch or loop) and the attempts made by irrigators to close the valve when it is not operating.

Measurements and visual estimates of outlet seepage losses, on systems in Bangladesh, are in the range of 0.4 to 0.6 l/s per leaking outlet valve (MMI, 1990a and Rashid et al, 1990). Total losses from structures on monitored Bangladesh systems were less than 4 to 6 l/s which on a pump discharge of 40 to 60 l/s represents a maximum of 10% of the flow rate.

Pipeline seepage losses are much more variable and depend on the quality of construction (particularly for non-reinforced concrete pipe), the length of the pipeline connected to the operating outlet, and the losses occurring at the joints. Measured values of seepage losses from concrete pipelines in Bangladesh were generally in the range of 0.3 to 0.7 l/s per 100m of pipeline (Rashid et al, 1990 — these figures include losses through outlets on the test line).

The work completed by Rashid et al (1992) agreed with the more limited measurements conducted by Ray (MMI, 1990b), suggesting that losses in buried pipe systems were at best 5% and at worst 15% of losses measured on the typical earth channel systems they replaced.

Table 1.3 gives estimates of pipe system transit efficiency based on field data for concrete pipe systems in Bangladesh and a range of other pipe systems visited throughout south Asia. The estimates for uPVC pipe systems are inferred from the very low frequency of pipe leakage (IDTP, 1989), observations of leaking outlet valves, and experience with uPVC pipe materials in other applications.

Note should be taken of the disappointing performance of many lined channel systems, and similar variation could be expected in the transit efficiencies of pipe systems, depending on quality of construction and maintenance.

Table 1.3 **Typical pipeline transit efficiencies**

	Good	*Average*
Non-reinforced concrete	90%	80-85%
Rigid uPVC	95%	90%

Source: Rashid et al (1992) and field observation of south Asian systems

The quaternary distribution efficiency must also be included to estimate the overall distribution efficiency of the system. This will depend on the method of quaternary distribution and the outlet to field distance.

Measurements of earth channel losses in Bangladesh (Rashid et al, 1992 and MMI, 1990b) illustrate the sensitivity of the quaternary channel distribution efficiency to the distance from the outlet to the field. The advantages of using buried pipe can be effectively cancelled by long travel lengths of earth channel, and therefore it is recommended that distances from outlets to fields never exceed 200m and are preferably around 100m.

Reduced land take

Where a choice is being made on a new irrigation scheme between an open channel and a buried pipe distribution system, the land take of each can be estimated from the layouts. Rashid et al (1992) measured the actual areas occupied by channels on three tube-well irrigation systems, before and after installation of the pipe system. The saving ranged from 0.5 to 2.4% (average 1.2%) of the original gross command area. While this area may appear small (0.48 ha on a command area of 40 ha), it may be important to the farmers concerned and in some intensively settled areas with very high land values it can make a valuable saving (Campbell, 1984).

Of more significance than the actual cost-saving may be the avoidance of major right-of-way and land acquisition problems which can slow and severely hamper any irrigation development.

Lower cost

The comparative costs (capital, maintenance and operation) of buried pipe systems and open channel systems are discussed in detail in Chapter 7. Several of the major conclusions are set out below to indicate the nature and magnitude of the cost benefits available.

Capital

Costs of buried pipe systems built in Bangladesh, Nepal and Indonesia, using a range of pipe materials, are from two to three times the cost of earth channel systems with water control structures, but consistently less than any of the hard surface lining systems considered. On a unit length basis, and considering the value of land saved, pipe systems cost from 10% less than ferrocement lining in Indonesia (Republic of Indonesia, 1990) to 50% less than the cost of brick lining systems in Bangladesh and Nepal (Gisselquist, 1986a; and GDC, 1987).

When the higher distribution efficiency of pipe systems is taken into account so that the cost of the water supply (in this example a deep tube-well) is spread over a larger area then pipe system costs on a per hectare basis become comparable to earth channel systems and are 50% of those for brick-lined channel alternatives (GDC, 1987).

Maintenance costs

Practitioners report lower annual maintenance costs of buried pipe systems in general (Campbell, 1984), and of higher pressure pipe systems in particular (Pimley and Fischer, 1990), but no data have been identified.

Pipe materials and methods of manufacture

Low-pressure buried pipe distribution systems are made from a wide range of pipe materials including:

- precast or cast-in-place non-reinforced concrete pipe;
- thin-walled asbestos cement;
- plastic pipe including smooth-walled and thin-walled uPVC pipe materials.

Non-reinforced concrete pipe materials have been successfully used for buried pipe distribution systems since the early 1920s (Coles, 1991) and have been widely used for pipe systems in the USA (Pimley and Fischer, 1990). While non-reinforced concrete pipe is strong in compression, it has weaknesses when subjected to tensile forces. However, a study in the USA (Koluvek, 1970) found that reinforced concrete pipe was too expensive for irrigation distribution systems.

Non-reinforced concrete pipes are still widely used in low-pressure pipe distribution systems, usually for situations where operating heads do not exceed 10m and are normally less than 4m (Merriam, 1990). Non-reinforced concrete pipe can be manufactured by a wide range of simple techniques including:

- mechanized vertical moulding techniques;
- hand-rammed vertical moulding;
- machine-spinning;
- hand-spinning;
- cast-in-place manufacture.

Most problems have been reported with hand-rammed vertically moulded pipe and small-scale spinning of non-reinforced concrete pipe, but experience in Bangladesh has shown that good quality pipe can be manufactured reliably by spinning (Georgi, 1989 and MMI, 1990a).

There is scope for the development of small-scale industrial manufacturing of concrete pipe for buried pipe irrigation systems. For example Georgi (1989) and MMI (1992) describe small pipe factories set up in Bangladesh, comprising a spinning machine and a curing pond. Quality control of materials and procedures is important for producing satisfactory pipe. (Further information is provided in Appendix 1.). The more commonly used joint systems include mortar-jointed, plane-ended pipe (MMP, 1989b), tongue and groove pipe with a mortar joint (Merriam, 1990) and spigot and socket pipe with a mortar seal or rubber gasket (Koluvek, 1970). Indian, American and British standards provide general specifications, although for irrigation use the American Society of Agricultural Engineers' standard provides the most relevant recommendations (ASAE S261.7, 1989).

Until quite recently the main plastic pipe material used for irrigation was smooth-walled rigid uPVC pipe. More recently new thin-walled corrugated uPVC pipe materials have been used for low-pressure irrigation applications (Jain, 1989). Opportunities presented by these new pipe materials are discussed in more detail in Appendix 1.

Although asbestos cement pipe is claimed to have long life and low maintenance costs, as demonstrated by many years of field experience (Univ. Cal., 1977), few buried pipe systems have been identified as using asbestos cement. Plusje (1981), however, reports on a pipe system built in Bogra, Bangladesh, where construction difficulties, seepage problems from ill-fitting joints and high installation costs discouraged repetition of the use of asbestos cement pipe. Asbestos cement pipe has long been used for non-pressure sewerage and stormwater applications in the UK and Europe (Everite, 1987).

The pipe materials outlined above are detailed more fully along with relevant references in Appendix 1. Various low-pressure pipe materials are illustrated in the associated publication 'Photographs of Buried Pipelines for Surface Irrigation'.

Classification of buried pipe systems

Definition of low-, medium- and high-pressure pipe systems

The classification which follows is based on consideration of the pipeline design working pressure. The design working pressure is defined as the sum of the static and dynamic pressures within the pipe system. The static pressure depends on the topography and the outlet operating pressure. The dynamic pressure comprises friction losses in the pipeline and structures and an allowance for any transient pressure surges associated with such events as water hammer. The design working pressure will usually be a maximum at the head of a pump supply pipe system or at the extremity of a steep gravity pipe system.

Low-pressure systems

Low-pressure buried pipe systems are defined as systems where the maximum design working pressure is less than 10m. This will include the majority of pipe systems which distribute water for application by surface irrigation methods.

In most cases, however, the maximum allowable pipe pressure will be considerably less than 10m, depending on the pipe material used, the method of pipe manufacture and the pipe jointing system used during construction. Table 1.4 details maximum allowable pipe pressures for pipe materials commonly used in buried pipe systems, based on guidelines presented in specific standards and field experience from Bangladesh.

Medium pipe pressures

Medium-pressure buried pipe systems are defined as systems where the maximum design working pressure is between 10 and 20m. Within this group will be included most trickle irrigation (or "micro-irrigation") systems, where working pressures in excess of 10m are often required to overcome friction losses in long pipe networks, filtration facilities at the inlet and automatic valving.

High pipe pressures

High-pressure systems are defined as buried pipe systems where the maximum design working pressure is more than 20m. This will include most sprinkler irrigation, except medium-pressure mini-sprinkler systems.

Classification of low-pressure buried pipe systems

Low-pressure pipeline systems can be classified on the method of pressure control into closed, semi-closed and open systems, and on the origin of the supply pressure into gravity, pumped or mixed systems. This follows definitions and terms used by Merriam (1987b) and Baudequin et al (1990).

Table 1.4 **Recommended maximum pipe pressures**

Pipe material	*Recommended max operating pressure*	*Max allowance for surge (m)*
Concrete		
Cast-in-place concrete pipe	2-5m	included
Hand-spun plane ended concrete pipe	4-5m	included
Machine-spun tongue and groove concrete pipe	5-8m	included
Machine-tamped tongue and groove or spigot socket pipe	5-10m	included
uPVC		
Thin-walled uPVC pipe (estimate) 2.5kg/cm^2	7m	3m
Rigid smooth-walled uPVC pipe	<18m	8m

Source: ASAE S376.1 (1989) and ASAE S261.7 (1989) and pipe manufacturers data from Jain, 1991

Notes

Non-reinforced concrete

For non-reinforced concrete pipes, standard specifications (ASAE S261.7, 1989) detail that the maximum operating pressure including pressure surge should not exceed 25% of the hydrostatic test pressure for mortar-jointed pipe and 35% of the hydrostatic test pressure for rubber gasket-jointed pipe. Although non-reinforced concrete pipes routinely test successfully to pressures between 1 and 2 bar (kg/cm^2), the cement mortar joint will not withstand these pressures and hence system operating pressures need to be down-rated to take account of this (MMI, 1990a).

uPVC (unplasticized polyvinylchloride)

For low-pressure uPVC pipe materials, standard specifications (ASAE S376.1, 1989) detail that:

- the maximum surge pressure should not exceed 30% of the rated pipe pressure;
- the total of maximum operating pressure and pressure surge should not exceed the rated pipe pressure.

Standard manufacturers' data on pipe pressure ratings assume a water temperature of 23°C, and where higher temperature water is conveyed the pipe's design pressure should be derated. Typically for a temperature of 44°C this would mean a halving of the pressure rating of the pipe, though generally water significantly warmer than 23°C is uncommon.

Values for derating, applicable to some plastic pipe materials, are detailed in ASAE S376.1 (1989); however reference should also be made to literature from the particular pipe manufacturer. Maximum allowable pressures, given in Table 1.4, allow for pressure surge at no more than 30% of the rated pipe pressure.

Care should particularly be taken when specifying thin-walled uPVC pipe materials for low-pressure applications, as only a few manufacturers currently specify a pressure rating for their pipe (Jain, 1991). Most thin-walled and corrugated pipe materials are recommended only for non-pressure applications (HPPA, 1989). New tests and specifications are considered necessary if these materials are to be used with confidence in low-pressure applications.

Method of pressure control

General

The concepts of open, semi-closed and closed systems have been introduced above. In a similar way to open channel systems, low-pressure buried pipe systems can be used in both flat and steep command areas, provided some mechanism for dissipating excess pressure head is provided to keep design working pressures to less than 10m, and maximum pipeline velocities to less than 1.5m/s. Two different methods of dissipating excess pressure are used in low-pressure pipelines.

In open pipe systems, overflow stand-pipes, which function like drop structures on open channels, are provided to dissipate excess head. Head is lost when water falls over the baffle in the stand to a lower level on the downstream side. Alternatively in semi-closed pipelines, the overflow stands are replaced with float valves, with the pressure reduction achieved by friction loss through the float valve.

Where no excess pressure is dissipated in these ways, the pipe system is termed closed.

The method of pressure control strongly influences the regulation of the pipe system, and this is discussed in detail in Chapter 2.

Open pipe systems

By utilizing overflow stand-pipes to dissipate excess pressure head, open pipe systems divide the pipeline into two or more hydraulically disconnected pipe sections with discontinuous hydraulic grade lines, and overflow stand-pipes acting as the boundaries. Because operating pressure cannot be transmitted between the different pipe sections, open systems are difficult to regulate and differences between inflow at the head of the system, and discharge from an outlet on a lower pipe section, result in operational spillage from the overflow stand between them.

Pimley and Fischer (1990) describe the problems associated with the operation of open pipe systems, and in particular the need for continual adjustment to balance inflow and outlet discharges to reduce operational spillage. Merriam (1987c) strongly advocates the conversion of existing open pipe systems to semi-closed ones to improve their operating characteristics, save labour and permit upgrading of operating schedules. Open pipe systems which are still in operation are described by Merriam (1987a) in the USA and Baudequin et al (1990) in France. A schematic diagram of an open pipe system is shown in Figure 1.4 b (page 7).

Semi-closed pipe systems

In semi-closed pipe systems float valves are used instead of overflow stand-pipes to dissipate excess head on sloping command areas. The float valves not only allow the whole pipeline to be hydraulically connected but also enable higher-pressure-rated pipe to be used, often reducing the required number of pressure control structures. Merriam (1987b) considers that semi-closed pipe systems combine the best features of both open and closed systems. The ability of closed systems to respond automatically to flow and pressure changes downstream is retained, so reducing the operational spillage and complex operation. The low operating pressures of open pipe systems are also retained, allowing the use of lower-cost pipe and equipment. This avoids the disadvantage of large pressure variations in closed systems where flow rates are changing.

Pipe systems of this type are little documented, and are presumably not widespread. One reason may be the relative obscurity of the design and performance of the float valve. Construction and operation of semi-closed pipe systems is described in Sri Lanka (Merriam 1985), the USA (Merriam, 1987a) and India (Merriam, 1990).

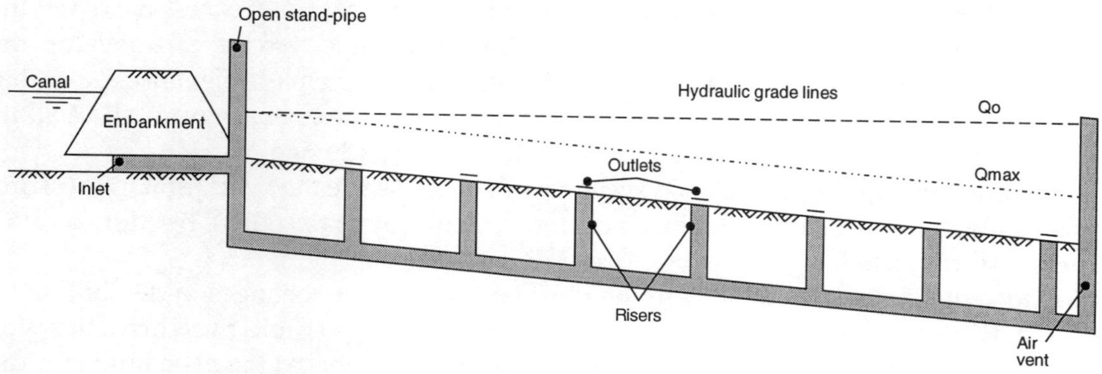

(i) Closed system — gravity

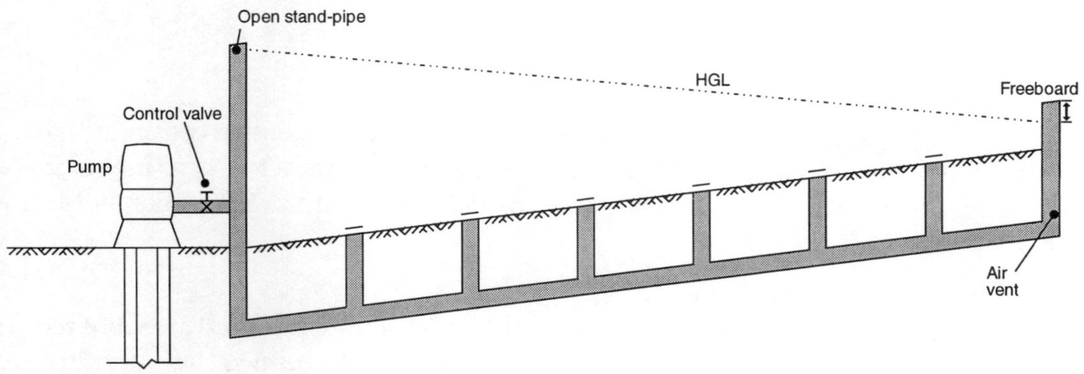

(ii) Closed system — pumped

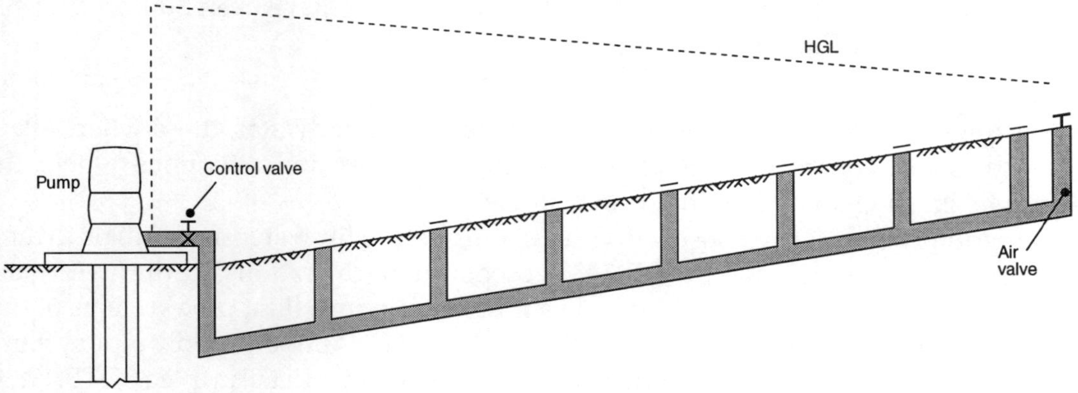

(iii) Closed system — pumped (with air valves)

Figure 1.6. Closed buried pipe systems

Source: Baudequin et al (1990)

Closed pipe systems

In a closed pipe system there is no requirement for the dissipation of excess head and the entire pipeline is hydraulically interconnected. Closed pipe systems can develop the maximum pressure resulting either from pumping or the difference in elevation. Note that closed pipe systems can be open to the atmosphere where the operating pressures allow above-ground structures to extend above the hydraulic grade line.

Merriam (1987c) notes that closed systems will have the greatest variation in pressure, so that stable pressure and flow rates become difficult to obtain where freedom to alter outlets is unrestricted.

Because these systems, which are the most common, allow changes in demand to be sensed at the water source, automation of the water supply is possible. Systems of this type are widely reported throughout the world with large numbers having been built in India (World Bank 1983), Bangladesh (Gisselquist 1986a, 1989) and China (DST, 1990).

Schematic illustrations of closed pipe systems under both gravity and pumped supply are shown in Figure 1.6.

Origin of the driving head

Gravity supply

If the available head is adequate to distribute water over the entire command area by gravity, then the system can be termed gravity supply. Systems of this type in the USA are described by Merriam (1987a) and Baudequin et al (1990), and in Sri Lanka by Merriam (1985).

Mixed gravity and pumped supply

Where the available head allows gravity supply to part of the command area, but requires pumping to the remainder, a mixed system can be used to reduce the pumping costs involved with a total pumped supply. Care must be taken to ensure that the two different systems are compatible in terms of control and operation.

Where the gravity portion of the scheme supplies water to the pumped section, a closed or semi-closed pipeline can be provided to enable automation of the pump and demand sensitive operation of the entire system. Systems of this type have been built in the USA and are described by Baudequin et al (1990).

Pumped supply

Systems under pumped supply include those utilizing groundwater, those where available head is inadequate to overcome pipe friction losses and situations where the topography is rising away from the water source.

In situations where farmers use individual pump sets to lift water from canals to their field channels such as in Egypt (EWUP, 1983), pipe systems can rationalize the investment in pumps, by enabling one pump to supply a number of farms. Pilot pipe systems of this type in Egypt are described by Hydraulics Research (1990c). Pumped supply pipe systems from tube-wells are widespread in south Asia, and are described in detail in India (IDTP, 1989), Bangladesh (MMP, 1989a), Sri Lanka (Merriam, 1985) and Thailand (HHP/HTS, 1988).

CHAPTER 2

DISTRIBUTION SYSTEM SELECTION

Introduction

This section details the procedure for selection of the buried pipe system and its components, which comprises the following steps.

- **Establishing the suitability of a buried pipe system over other types of distribution system**

  Buried pipe systems will not always be the most appropriate choice. Given information on the area to be irrigated, water source, irrigation opportunities and social and environmental factors, it should be possible to establish whether a buried pipe system is an attractive option.

  Suitability is influenced most strongly by issues of soil texture, topography of the command area and water availability. This is discussed in detail below and summarized in Flow chart 2.1. The cropping pattern is also important, and both diversified cropping systems and high value crops will favour the use of a buried pipe system with greater operational flexibility than an open channel system.

- **Selection of the type of buried pipe system best suited to the topography and water source available**

  Selection is based on questions of the hydraulic operation of the system (see Flow chart 2.2) and the source of the driving head for the system (see Flow chart 2.3). Reference should be made to the definitions of pipe systems contained in Chapter 1. For each situation there may be more than one alternative and then choice should take into account other factors such as cost and management flexibility.

- **Selection of the method of regulating water inflow to the pipe system**

  Regulation refers to the way in which matching of supply with demand is achieved. The alternative methods of regulation available depend on the type of hydraulic operation of the pipe system and whether the supply is from gravity or pump.

- **Selection of the operating and control system to be used**

  Once the choice of regulation has been made a choice needs to be made of the operating and control system. In some cases the selected method of regulation will leave only one way of operating and controlling the system.

- **Selection of the pipe layout and pipe material to be used**

  At this stage initial selection is made on the basis of technical feasibility and total cost given that the pipe is normally the largest proportion of the system cost. Where cost of different pipe materials is comparable, other considerations such as ease of construction become important. During later design, a final choice will be made of the pipe sizes and jointing method to be used.

Suitability of buried pipe systems

Alternative distribution systems

For water distribution within the tertiary unit of an irrigation scheme various alternatives are available, including:

- earth channel (preferably with compacted banks);
- various in situ and prefabricated concrete canal linings;
- brick-lined canals;
- low-cost membrane lining, including plastic, asbestos and rubber materials;
- low-pressure pipe systems of concrete, uPVC or asbestos cement;
- mixed systems with differing proportions of any of the above.

All systems should include appropriate water control structures.

The qualitative and quantifiable benefits of buried pipe systems have been described in Chapter 1. While the performance of open channel systems is often inferior to that of pipe systems, it is only in particular circumstances that pipe systems are clearly preferable.

The distribution systems considered here are assumed to comprise extensive networks with from 1000 to 2000m of channel or pipe. For very short networks (small command areas) earth channel will frequently still be the most economical choice.

In general, wherever lining is considered to be justified, a buried pipe system should be a lower-cost alternative with superior performance, longer life and operational advantages, provided it is constructed to an adequate standard.

The most attractive situations for a buried pipe system are where it can be operated under gravity supply, or where pumping would be required for an open channel system and the additional head required for the buried pipe system does not greatly increase costs or reduce the discharge.

Less attractive are schemes with existing gravity-fed open channel distribution systems, which would require a pumped supply for a buried pipe system.

The situations and circumstances in which pipe systems are considered attractive relate particularly to questions of soil texture, topography of the command area and water availability.

Flow chart 2.1 summarizes the main criteria used in selecting pipe systems ahead of open channel alternatives, while Flow charts 2.2 and 2.3 detail selection of the type of pipe systems suited to each situation. Buried pipe distribution systems are particularly suitable where there are coarse-textured soils, or valuable and limited water supplies, or topographic problems in the command area, as explained below.

Coarse-textured soils

Earth channel systems built in sandy soils with poor cohesion will have potentially high seepage losses, require frequent and heavy maintenance, and will often have a very limited life. Such soils also appear to be subject to a higher level of animal burrowing activity, and root penetration. The high transit efficiency of buried pipe systems is a major advantage in this case.

Canals built with coarse-textured soils will also often require flatter embankment side slopes, so taking up more land for the channel network.

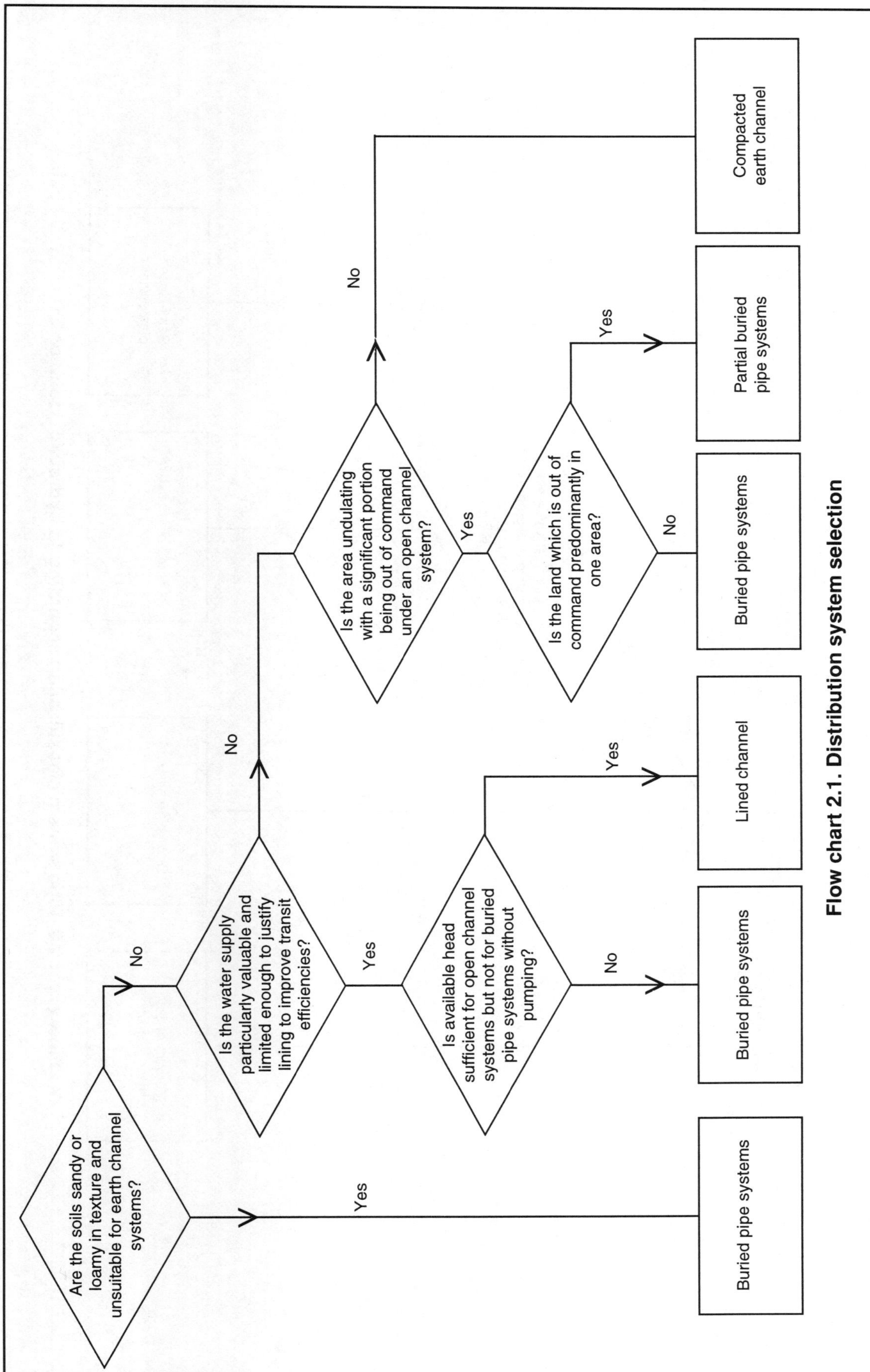

Flow chart 2.1. Distribution system selection

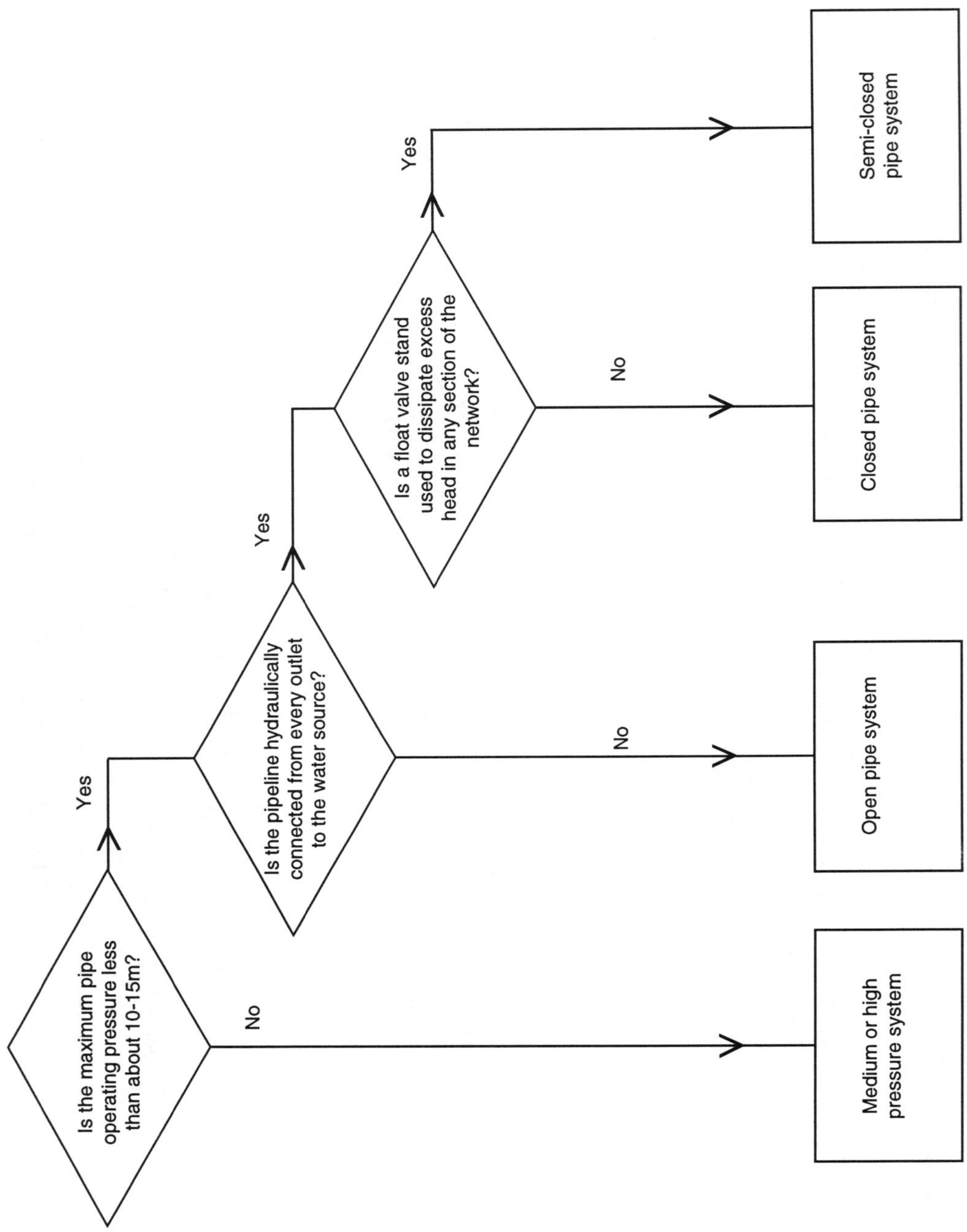

Flow chart 2.2. Pipe system classification according to hydraulic operation

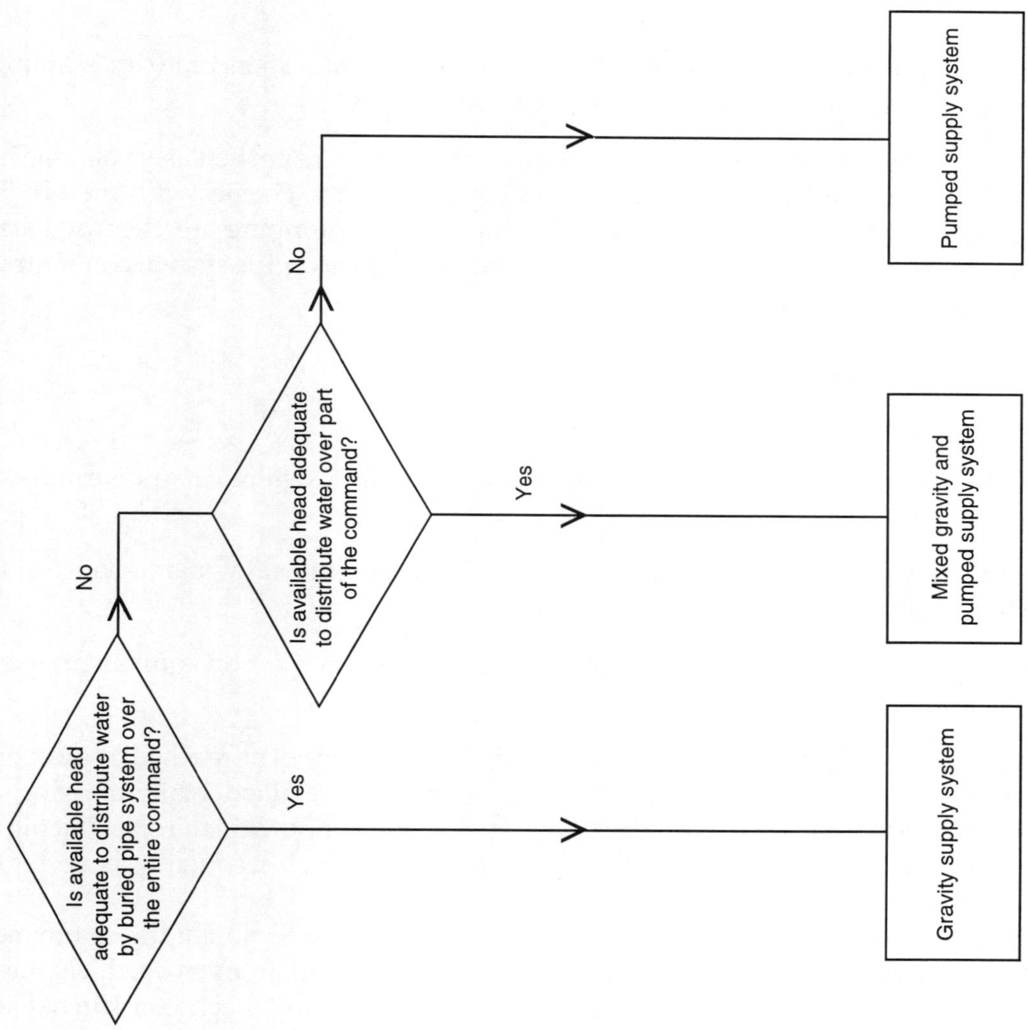

Flow chart 2.3. Pipe system selection according to origin of pressure

Valuable and limited water source

Buried pipe systems are particularly attractive for their high efficiency and flexible operation where water is both valuable in terms of the crops which can be grown, and limited in availability.

Examples of the use of buried pipe distribution systems in these circumstances include:

- water-harvesting reservoirs where supplemental irrigation of rainfed crops can provide dramatic yield responses, and water demand greatly exceeds supply (schemes in northern India described by Seckler, 1986);
- in regions of poor groundwater where abstraction from surface water sources during the dry season is limited (observed on pump lift irrigation systems visited in Maharashtra, India);
- where irrigation of high-value crops can provide significant cash income (for example in Tangail, Bangladesh, as described by Gisselquist, 1989).

The higher distribution efficiencies of buried pipe systems are particularly valuable in the case of deep groundwater abstraction and high pump lift schemes with their high pumping costs. The limited water supply and high cost of pumping are cited by local irrigation practitioners as reasons for the extensive use of buried pipe systems on central Indian pump lift irrigation schemes.

Command area topography

Pipe systems are particularly attractive where:

- they allow irrigation of land out of command under an existing canal distribution system;
- they replace sections of raised earth channel which require heavy maintenance and have high seepage losses;
- it would be difficult or expensive to obtain the land required for an equivalent open channel system.

Where variations in ground level mean that irrigable land cannot be reached by an open canal system, a pipe system allows hydraulic head to be retained while crossing a depression or cross-drainage. Short lengths of pipeline may be sufficient, replacing high embanked channels which are difficult to maintain, and which frequently have high seepage losses.

Where an area of high or undulating land is confined to one section of the command area, an attractive solution is a partial buried pipe system with an open earth channel system in the remainder of the area. Systems of this type have been constructed in Bangladesh (MMI, 1990a) and India (MMP, 1989a), and have received ready acceptance by irrigators. They have a lower capital cost than full pipe systems.

Whereas land for new canals may be relatively easy to obtain in previously uncultivated areas, this is much more difficult where land is already intensively used — for example on a project to improve an existing irrigation system.

The delays and costs associated with obtaining land for an open channel distribution system may be reduced by using a pipe system with its lower land requirements, as found in Uttar Pradesh and Kerala, India (Kolavalli and Shah, 1989, and Campbell, 1986).

Methods of regulating inflow to pipe distribution systems

All closed and semi-closed pipe systems can be operated on demand. This means that any alteration in flow requirements downstream can be sensed either by a pump or gate operator or by upstream level control equipment or flow sensors. The choice for regulating inflow to the pipe system is between manual and automatic methods.

The choice of regulation method is important on all pumped supply systems and will be between a manual or automated system. In all situations several methods of regulation will be available with manual regulation always an option.

Regulation of inflow to open pipe systems

The inflow to all open pipe systems must be regulated manually, if excessive operational spillage at the overflow stands is to be avoided. This involves careful setting and resetting of control valves within the system. Automatic control is only possible if the system is converted to a semi-closed one by the installation of float valves in the overflow stands. Open systems with manual regulation are found in western USA.

Regulation of gravity inflow to semi-closed and closed pipe systems

In gravity-supplied semi-closed and closed pipe systems the inflow to the pipe system can be self-regulating, provided structures are installed on the pipeline to cope with pressure variations created by sudden changes in flow. The inflow to the pipe system may still be restricted by the use of gates for operational reasons. As long as the inlet structure to the pipe system extends above the canal or reservoir water level, spillage will not occur.

Where desired, control of inflow to the pipeline from canals with varying water levels can be provided by a range of structures including constant water level control gates. Otherwise inflow will be set by the maximum flow capacity of the pipeline determined during design.

Regulation of pumped inflow to semi-closed and closed pipe systems

The choice of method of regulation will depend on the method used to balance the pump discharge and outlet demand. The choices vary according to whether the pump is driven by an electric motor or a diesel engine.

Methods of balancing pump supply and demand

Four distinct methods of balancing inflow and outflow from a pumped supply pipe system can be identified.

- **Altering the pump speed and hence discharge to match the irrigation demand more closely**
  With diesel engines it is often possible to vary the engine speed in the range between nominal speed and 50% below, generally with a small reduction in efficiency. This practice is observed on tube-well-based systems in Bangladesh, although reduced engine speed is often due to poor engine installation or maintenance.

  Electrically-driven pumps generally have fixed speeds of 1500 and 3000 rpm based on fixed electric motor speeds. Electric motor-driven pumps can, however, be varied in two ways:

  i) by varying the frequency of the input power supply directly or

  ii) by using diesel engine generators to drive the electric motors and to vary the speed of the generator to produce the power frequency required.

Varying the speed of electric motor-driven pumps has not been used on any of the buried pipe systems studied. It is more expensive than fixed motor speeds and is unlikely to be appropriate for many developing country situations.

- **Altering the number or setting of the outlets which are operating, to suit the pump discharge**

  Most carefully designed small pipe systems can balance pump inflow and outflow demand in this way once the operator has sufficient experience. This method is commonly used where pumps are directly connected to the pipe system. Most of the pipe systems identified in China and Indonesia rely on this method.

- **Alternate switching of the pump off and on, either manually or automatically, to achieve a desired line pressure or water level in a control reservoir**

  Automated methods of regulating electric pumps usually rely on water level or line pressure sensing. Storage in a pressure vessel or tank must be adequate to allow for normal operation without excessive pump cycling which would lead to failure of electrical or mechanical components. Diesel engine-driven pumps should not be stopped and started frequently.

- **Selecting one or more pumps from a set, to match the demand flow rate**

  This method is restricted to larger schemes with a wide range in possible operating demand which justifies the investment in a set of pumps. This system has been used on a pipe system in Egypt and is described by EWUP (1983).

Systems of regulation for pumped inflow

By considering the method of balancing inflow to the pipe system with the requirement for automation, a system of regulation can be defined and the alternatives are described below.

For electric motor-driven pumps an automatic system of regulation is possible, while for diesel engine-driven pumps only manual methods are possible. If the diesel engine-driven pump is directly coupled to the pipe system (for medium pressures and higher flow velocities) then only manual regulation is practical, by altering pump speed and outlet settings. If there is a header tank then manual regulation can also be carried out by observing the water level in the tank.

For electric motor-driven pumps, where the irrigation demand is large, widely variable and provided from a surface water source, several alternatives exist including:

- manual regulation of multiple pumps;
- variable speed motors;
- automatic regulation of multiple pumps by flowmeter.

For other electric motor-driven pumps, selection will depend on whether the pipe system is directly connected to the pump, or flow is to be divided between several separate pipe networks at the pump.

If an elevated tank or header tank is provided then regulation can be manual or automated using float switches or probe controls within the tank.

The choice of the method of pump regulation needs to take into account the skill level of the irrigators who are to operate and maintain it, and the resources available for carrying out any repairs. Wherever automatic regulation is provided, provision should be made for manual regulation, with necessary protection, in the event of failure of the automated system. In most developing countries the difficulties of maintaining automated systems, obtaining spare parts and the need for flexibility in supply will mean that systems of manual regulation are frequently preferred.

Labye et al (1988) define and describe alternative manual and automated methods of pump regulation for buried pipe systems. These are summarized below and discussed in more detail in Chapter 3.

Manual systems

- **Simple method**

  Pump is controlled by the operator and is either off or on: typical of systems where the pump is directly connected to the pipe system.

- **Water level observation**

  Pump is controlled by the operator to maintain a desired water level or line pressure.

Automatic systems

- **Regulation using an elevated open tank or reservoir**

  Two methods are used:

  i) Pressure sensor regulation

  A pressure sensor on the pump delivery pipe switches the pump off when the line pressure exceeds a certain value, set by the closing of a float valve on the pipe entry to the elevated tank: suited only to single pump installation on small irrigation schemes.

  ii) Water level method

  The pump is regulated by the use of float switches or probe controls in response to the water level in the elevated tank. Suitable where the regulating tank can be sited close to the pump (illustrated in the associated publication 'Photographs of Buried Pipelines for Surface Irrigation').

- **Air pressure vessel method**

  Pump is controlled by pressure sensors operating over a range of pressures for which flow is attenuated in a pressure vessel: suitable for medium- rather than low-pressure pipe systems.

- **Regulation by flowmeter**

  A number of pumps are switched on or off in response to flow measured through a flowmeter: suitable on larger buried pipe systems utilizing a series of pumps to satisfy widely varying flow demand.

In all closed and semi-closed pipe systems water level changes back at the inlet or pump will reflect changes in demand and either automatic or manual adjustments can be made to the irrigation supply. In an Egyptian-built pipe system excess water which cannot be used spills back into the pump sump or well (EWUP, 1983). Although this represents an operating inefficiency, it may be the most practical way of utilizing fixed discharge pumps while satisfying a widely varying demand.

Selection of operating and control system at the head of the pipe system

The operating and control system comprises the method of regulating inflow to the pipe system and the type of control structures at the inlet and head of the pipe system. The choice of operation and control requires a design decision and a choice of technology. The appropriate pipe system design, in terms of operation and control, will depend on such factors as:

- physical features of the command area;
- irrigator expectations of system performance including equity, variability in system discharge;
- irrigation methods practised.

This should be distinguished from the mode of operation, explained in the following section, which relates to the choice of how water is to be distributed within a particular design of pipe system.

In the following discussion the inflow to the pipe system from the pump or other water source is termed the inflow discharge or pump discharge (Q_i), flow from an outlet is termed outlet discharge (Q_o), and the flow in the quaternary field channel is called the field discharge (Q_f).

Whether the supply is gravity or pumped will define the range of possible operating and control regimes available. For each operating and control system there is a design mode of operation, as well as a range of other possible modes of operation. Irrigator preference should be considered when making the choice. Depending on the selection and arrangement of control structures within the system some modes of operation will not be practical.

Flow charts 2.4 and 2.5 summarize the criteria, which are also listed below, for selection of the operating system for pumped and gravity supply systems respectively.

For pumped systems important issues include:

- the ratio of the design outlet discharge (Q_o) to the design inflow discharge (Q_i) and hence the number of outlets to be operated at one time;
- how pump operation is to be regulated;
- whether accurate division of flow is desired, and if so whether this should take place at the water source, within the pipe system or at the outlet.

For gravity systems additional criteria include:

- whether the available head exceeds the desired maximum pipe-operating head in any part of the command area;
- whether on-demand operation of the system is desired;
- whether manual or automatic water level control at the inlet is desired.

For both gravity- and pumped-supplied systems the range of operating and control systems most commonly found is outlined below. This is by no means exhaustive and other alternatives will exist. At a first reading some may wish to skip this section and return to it later.

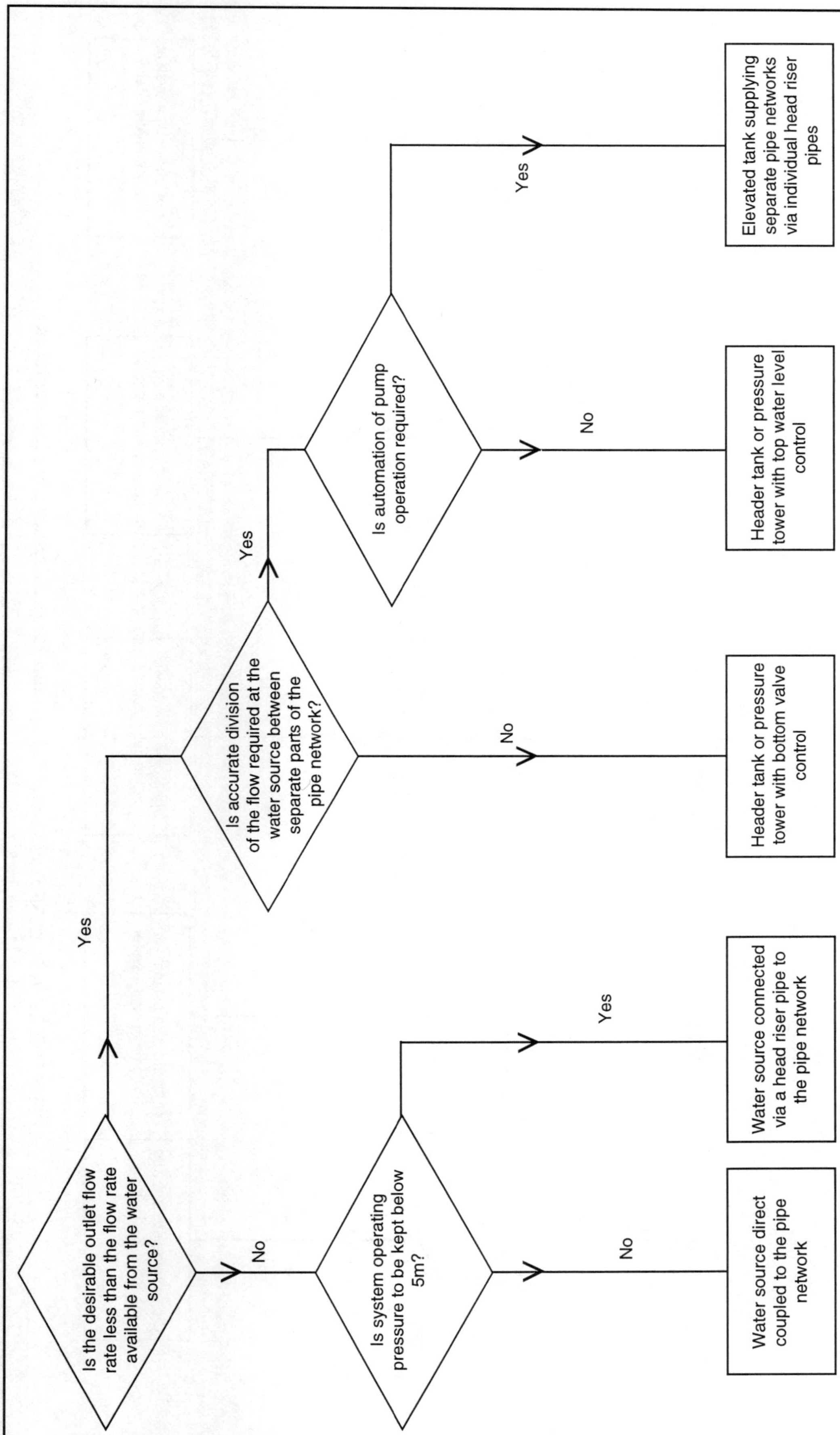

Flow chart 2.4. Selection of operating system for pumped-supply semi-closed or closed pipe systems

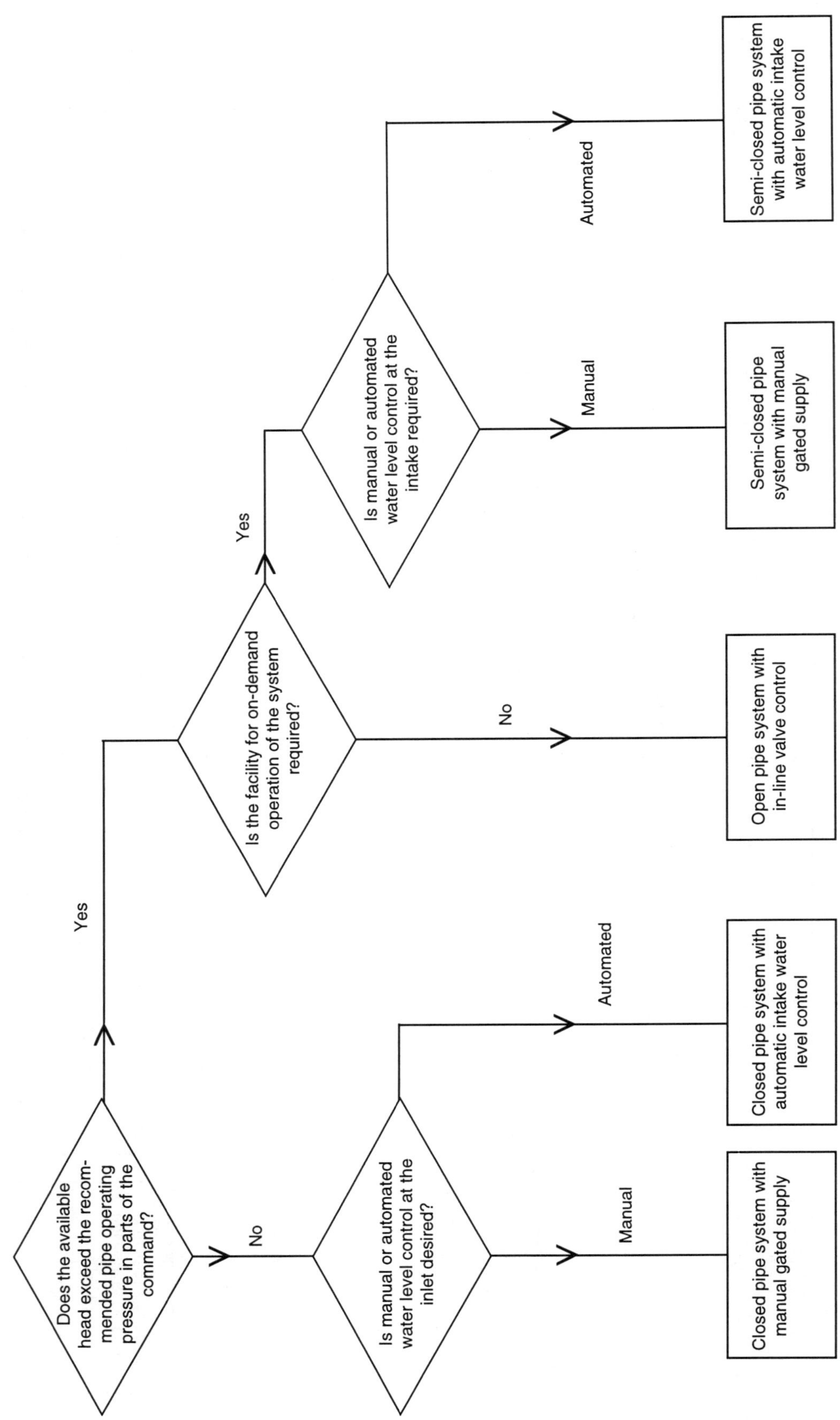

Flow chart 2.5. Selection of operating system for gravity-supply pipe systems

Pumped inflow systems

❍ **Pump connected directly to the pipe network** (Figure 2.1)

In this case the typical system operating pressure is medium (> 10m) rather than low as higher pressures and flow velocities are accepted to try to reduce pipe diameters and cost. Either a pressure release valve or pressure sensors to switch off the pump are provided to protect the pipe system from excessive line pressures. Generally manual operation is practised as automation is only practical on very small electric pump supply systems using a pressure vessel and preset pressure sensors. Systems of this type have been built in Thailand and are described by HHP/HTS (1988).

❍ **Pump connected via a head riser pipe to the pipe network** (Figure 2.1)

Because water at the inlet to the pipe system can rise no higher than the practical height of the head riser pipe, operating pressure is typically restricted to less than 5m, thus avoiding the need for special line protection. Systems of this type are described as having been built in Indonesia, China and Bangladesh.

❍ **Header tank or pressure tower with distribution by bottom valve control** (Figure 2.1)

In this system water is discharged into a vertical large diameter open stand, called a pressure tower or header tank. The diameter is such as to enable inlets to two or more pipe networks to be installed in the floor of the tank. This allows supply to be distributed to one or more parts of the system while shutting off those sections which are not operating. This reduces the losses from unused pipe branches due to seepage which would occur on non-reinforced concrete pipelines. Matching of inflow discharge and outlet discharge is usually regulated manually, by varying engine speed in the case of diesel motor installations or by altering outlet valve positions and settings with electric motor-driven pumps.

The volume of the header tank is not usually large enough to allow for automation by probe control without risking excessive pump cycling. Systems of this type have been built in Bangladesh and India, and are described by MMP (1989d, 1989e), Georgi (1989) and Gisselquist et al (1986a).

❍ **Header tank or pressure tower with top water division** (Figure 2.1)

In this system the inflow to the pipe system enters separate pipe networks via satellite stand-pipes. Water is drawn from the upper portion of the header tank or pressure tower flowing into two or more satellite stands which may be fitted externally or internally. The system allows for inflow to the pipe system to be divided in fixed proportions between two or more pipe systems. One major advantage with this system is that the flow is reduced before entry into the buried pipe system and so a reduction is possible in the pipe sizes used, with a resulting cost-saving.

In Bangladesh, internal or external satellite towers are used. The internal satellite stands include provision for shutting the supply to any of the networks by closing an alfalfa valve.

A number of header tank designs which have been built in Bangladesh on the IDA DTW II and TADP Projects are illustrated in the associated publication 'Photographs of Buried Pipelines for Surface Irrigation'.

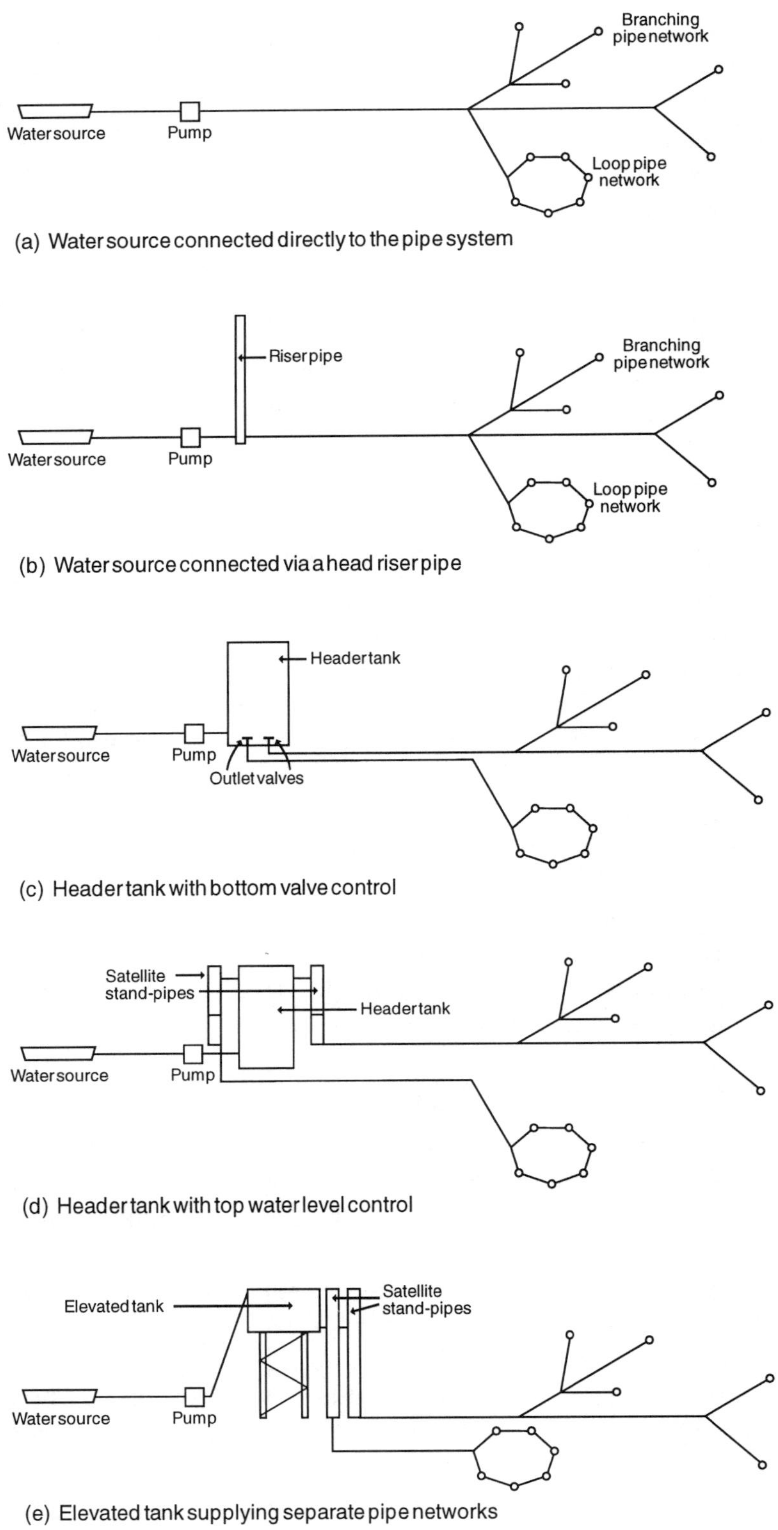

Figure 2.1. Schematic illustration of pumped-supply operating systems

A variation on the use of satellite towers is the provision of outer segments within the design of the header tank. Weirs set between the inner section and the outer chambers, connected to separate pipe networks, control the flow of water to each pipe system. In-line valves installed at the head of each pipe network allow for flow regulation and shutoff.

This system is used in the header tanks built in the pump lift irrigation schemes in Maharashtra, India.

- **Elevated tank supplying separate pipe networks via individual stand-pipes** (Figure 2.1).

In this system water is delivered by pump to an elevated tank. Inflow to the pipe system is via vertical stand-pipes connected to separate pipe networks. As with the previous system, the reduction in the inflow discharge to each pipe network allows for a reduction in pipe diameter and resulting cost-saving. These systems can be automatically or manually regulated.

Automation is achieved using probe controls to switch electric motor-driven pumps off and on, and systems of this type have been built and are still being constructed in Uttar Pradesh, India. The elevated tank illustrated in the associated publication 'photographs of Buried Pipelines for Surface Irrigation' has a probe control for automatic regulation installed. The systems are also described in more detail in project documents including World Bank (1983) and IDTP (1989). Elevated tanks of a similar design have been built under the Bhairawa Lumbini project in southern Nepal, although they allow for flow division between up to four different pipe networks by the use of stop logs.

Gravity inflow systems

- **Semi-closed or closed pipe system with manually gated inflow**

The gate allows for shutting off the inflow to the pipe system when water is not required or reducing the maximum inflow to something below that set by the design capacity of the pipeline and the size of the riser and valve opening. Provided water is available, inflow to the pipe system will automatically balance the combined outlet discharge.

Systems of this type drawing water from major canals have been in use for many years in California and other western states of the USA. The Gadigaltar scheme in Madhya Pradesh, India, is a more recently constructed example of this type of system and a typical screened gravity inlet with gate is illustrated in the associated publication 'Photographs of Buried Pipelines for Surface Irrigation'.

- **Semi-closed or closed pipe system with automatic water level control from the intake**

In a situation where the level of the water source fluctuates it may be desirable to install a structure to maintain a controlling water level height at the head of the pipe system. This will usually involve the installation of automatic water level sensing gates or devices on the intakes to pipe systems. Provided water is available, the system can be operated on demand. Individual outlet discharges (Q_o) will still be influenced by the number and position of irrigators taking water but the requirement for labour is reduced. Installation of this type of control, together with the upgrading of open pipe systems to semi-closed, has been a feature of recent improvements to buried pipe distribution systems in the USA (Merriam, 1987a).

❍ **Open pipe system with in-line valve control**

On open pipe systems, the intake, inflow discharge and the combined outlet discharge must be balanced. Repeated adjustments, which are time-consuming, need to be made to balance flows and avoid excessive spillage at the overflow stands. These must be repeated whenever changes are made in outlet valve settings . Pimley and Fischer (1990) describe the difficulties of this type of operation and the labour-intensive nature of the work.

Modes of operation for irrigation distribution from buried pipe systems

Mode of operation is used to describe the way in which distribution of water occurs within a pipe system. For any pipe system, there is more than one possible mode of operation, though it is likely that the designer will have intended one mode to predominate.

The mode of operation depends on the ratio of the design inflow discharge (Q_i) to the field discharge which can be used efficiently by each farmer (Q_f). In most small basin irrigation schemes this flow (Q_f) is of the order of 20-30 l/s for earth channels at the quaternary level. Thus where the inflow to the system is greater than 20-30 l/s then some method of flow division needs to be used.

For the same pipe system, as irrigator requirements for duration and rate of flow vary so the choice of mode of operation may change. The choices available will in part be determined by the selection and arrangement of structures within the system.

Some of the main issues are summarized in Flow chart 2.6. The mode of operation will have some influence on the adequacy and reliability of the supply delivered to the outlet.

Possible modes of operation are described below.

Mode A

One irrigator takes the full inflow discharge at one outlet ($Q_i \approx Q_o \approx Q_f$, depending on losses).

For any pipe system where the full inflow discharge can be handled efficiently by the irrigator, this mode of operation should be used. (That is, for Q_i not more than about 30 l/s for earth channels at the quaternary level.) Typically small buried pipe systems where the pump is directly connected or connected via a head riser pipe should be operated in this way.

All the remaining modes of operation relate to systems where the inflow discharge is greater than the optimum field discharge.

Mode B

The full inflow discharge is delivered to one outlet, where it is shared between two or more farmers by means of an outlet distribution structure ($Q_i \approx Q_o \approx 2\,Q_f$, depending on losses).

This mode of operation provides an equitable division of supply but requires a flow division structure at the outlet. Experience in Bangladesh suggests that problems can be encountered where co-operation among the irrigators is weak, and during low demand periods when it is difficult to co-ordinate the requirements of two irrigators at one outlet who wish to take water simultaneously. This often results in irrigation by one irrigator using the undivided inflow discharge, Q_i, with consequent overtopping of earthen field channels and high field distribution losses.

Mode C

In this case the inflow discharge is divided at the water source between two or more separate pipe networks ($Q_i \approx n\,Q_o \approx n\,Q_f$, $n \geq 2$).

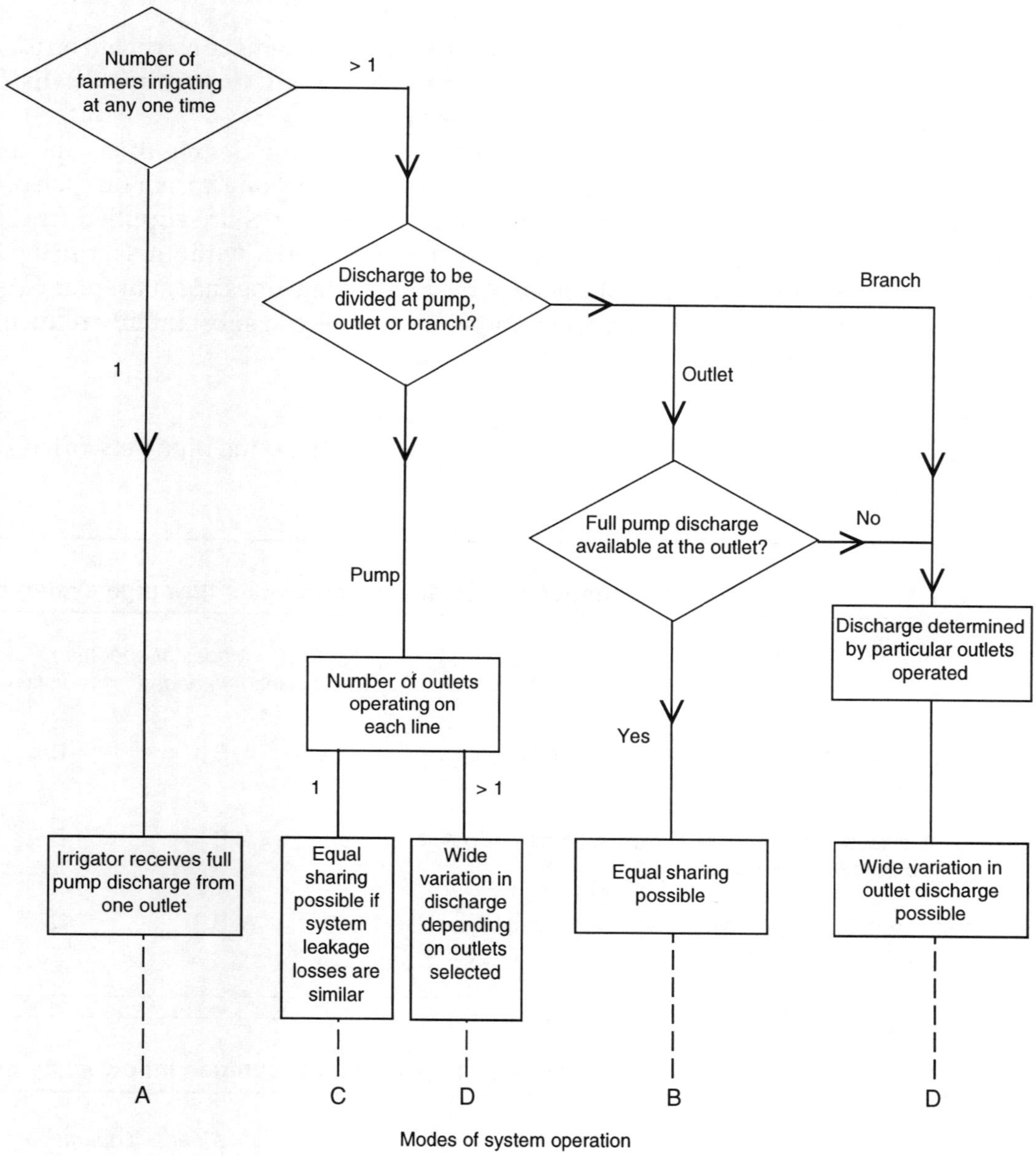

Flow chart 2.6. Alternative choices for operation of pipe systems

In this mode of operation, structures at the head of the pipe system allow the inflow discharge to be divided between different parts of the pipe system. Structures such as the elevated tank, and the header tank with top water level division to external satellite stands, result in a more equitable division of flow than the header tank with bottom valve control. Weirs and gates can also be used to divide the inflow discharge between different pipe networks according to their net irrigable command areas.

This mode of operation is provided for in the uPVC pipe systems constructed on a large scale in Uttar Pradesh, India. Where seepage losses are small, flow is evenly divided between two outlets, each operating on a different pipe loop.

This method is by far the most preferable and reduces the degree of co-operation necessary to the co-ordination of irrigation water use by at least one farmer on each of the pipe networks operated. In situations where four pipe networks are supplied from the same source, water can be delivered to two or three networks without sacrificing the principle of equity. Flow division at the head of the pipe system does not guarantee equity, as differences in leakage rates over varying pipeline lengths can substantially reduce the outlet discharge (MMI, 1990b).

Mode D

The inflow discharge is shared between several outlets on the same pipe network ($Q_i \approx n\,Q_o \approx n\,Q_f$, $n \geq 2$).

Table 2.1 Possible modes of operation for gravity inflow pipe systems

Operating systems	*Modes of operation* Recommended	*Modes of operation* Alternatives
1. Open pipe system with in-line valve control to each pipe network	A, B, C	D
2. Semi-closed or closed pipe system with manual gated inflow from reservoir or canal	A, B, C	D
3. Semi-closed or closed pipe system with automatic water level control at the inlet	A, B, C	D

Table 2.2 Possible modes of operation for pumped inflow systems

Operating systems	*Modes of operation* Design	*Modes of operation* Alternatives
1. Water source directly connected to the pipe network	A	D, B
2. Water source connected via a head riser pipe to the pipe network	A	D, B
3. Header tank or pressure tower with bottom valve control	D, A	B
4. Header tank or pressure tower with top water level control	C	D, B
5. Elevated tank supplying separate pipe networks each with an individual stand-pipe	C	D, B

Although all pipe systems can be operated in this way, it cannot be recommended because wide variations in outlet discharges can occur from differences in the heads at the outlets. Adjustments to flow can be made by throttling individual valves or using gate stands to divide water between branches. These changes are irregular and difficult to predict, and need to be repeated when a different combination of outlets is operated.

For both gravity and pumped supply pipe systems, the predominant design modes of operation are listed below. For each system alternative modes of operation are possible but have significant limitations.

Pipe system layouts

Buried pipe systems may comprise either a branch or loop layout, or a combination of both. Although in theory loop pipelines offer considerable advantages in terms of hydraulic performance they are not widely adopted for practical reasons outlined below.

Some of the issues involved in the selection of the pipe layout are summarized in Flow chart 2.7.

Branch layout

In a branch layout the pipelines may divide but they do not join up into loops. Water flows only in one direction, away from the water source, to outlets on each of the branches.

Branch systems require larger diameter pipes than loop systems because they must take the entire outlet discharge in one pipe. If the pipeline is prone to leakage then a branch system is recommended as parts of the system can be isolated when not in use, using valves or gates in stands at the head of each line. Maintenance and leakage repair is then facilitated by the ability to isolate particular parts of the system.

Some leakage is inevitable on concrete pipe systems and they usually have branch layouts, especially as reducing the pipe diameter has little effect on the cost of concrete pipe.

An analysis of idealized layouts shows that for the same command area a branch layout will typically have less pipe than a loop layout with a similar density of outlets and length of field channels (see Chapter 3). It is easier with a branch system to achieve an even density of outlets with acceptable distances from outlet to field plots, particularly where the command area is fragmented or irregular in shape. A typical branch pipe layout as designed by IDA DTW II Project in Bangladesh is shown in Figure 2.2 (MMP, 1989b).

Loop layouts

In loop layouts the flow is divided between the two halves of the loop so that water flows to an outlet by two routes. This allows about a 25% reduction in pipe diameter. Whether a loop system is cheaper in cost than a branch system depends on whether the saving due to a reduction in pipe size is more than the extra cost of the greater length of pipe.

Only closed pipe systems or closed parts of systems can utilize loop pipelines within their network. Open or semi-closed systems, however, can include loops as part of a closed pipe section, i.e. where both ends of the loop close on a single connected section of pipe and each part of the loop is hydraulically connected.

Because a loop pipeline cannot be isolated easily with valving without reducing the system capacity or complicating system operation, it is important that the rate of leakage from the pipe is low. When the loop remains full of water, flow will occur from the outlet almost immediately after the valve is opened and the water supply actuated. Both uPVC and asbestos cement pipe systems if well built have very low seepage losses and have been used for loop layouts. A schematic drawing of a uPVC loop system and header tank with bottom valve control, as built in Bangladesh, is shown in Figure 1.3. Figure 2.3 shows a double loop layout of a typical uPVC pipe system as built in Uttar Pradesh (Cunningham, 1986).

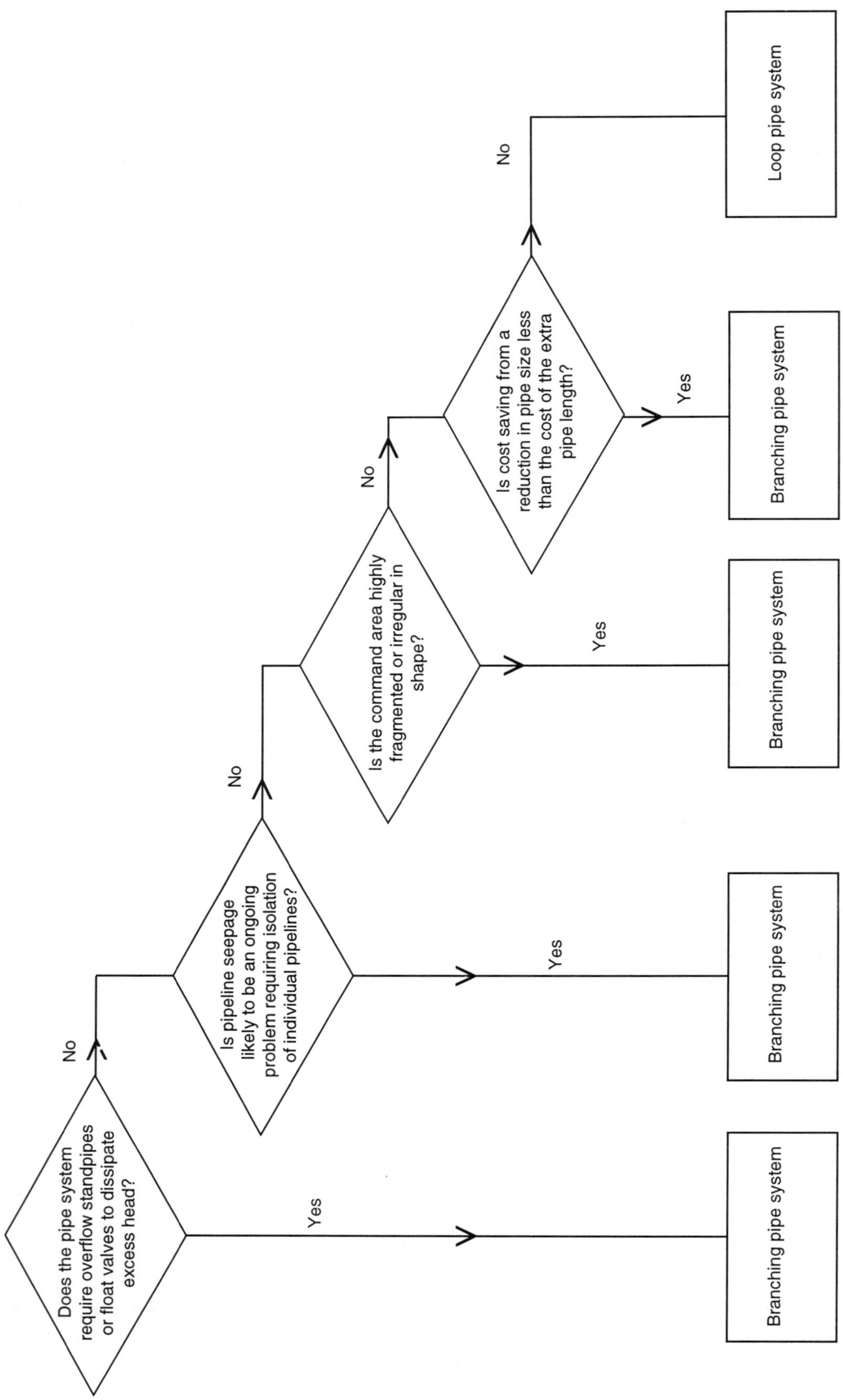

Flow chart 2.7. Selection of pipe layout: branching or loop

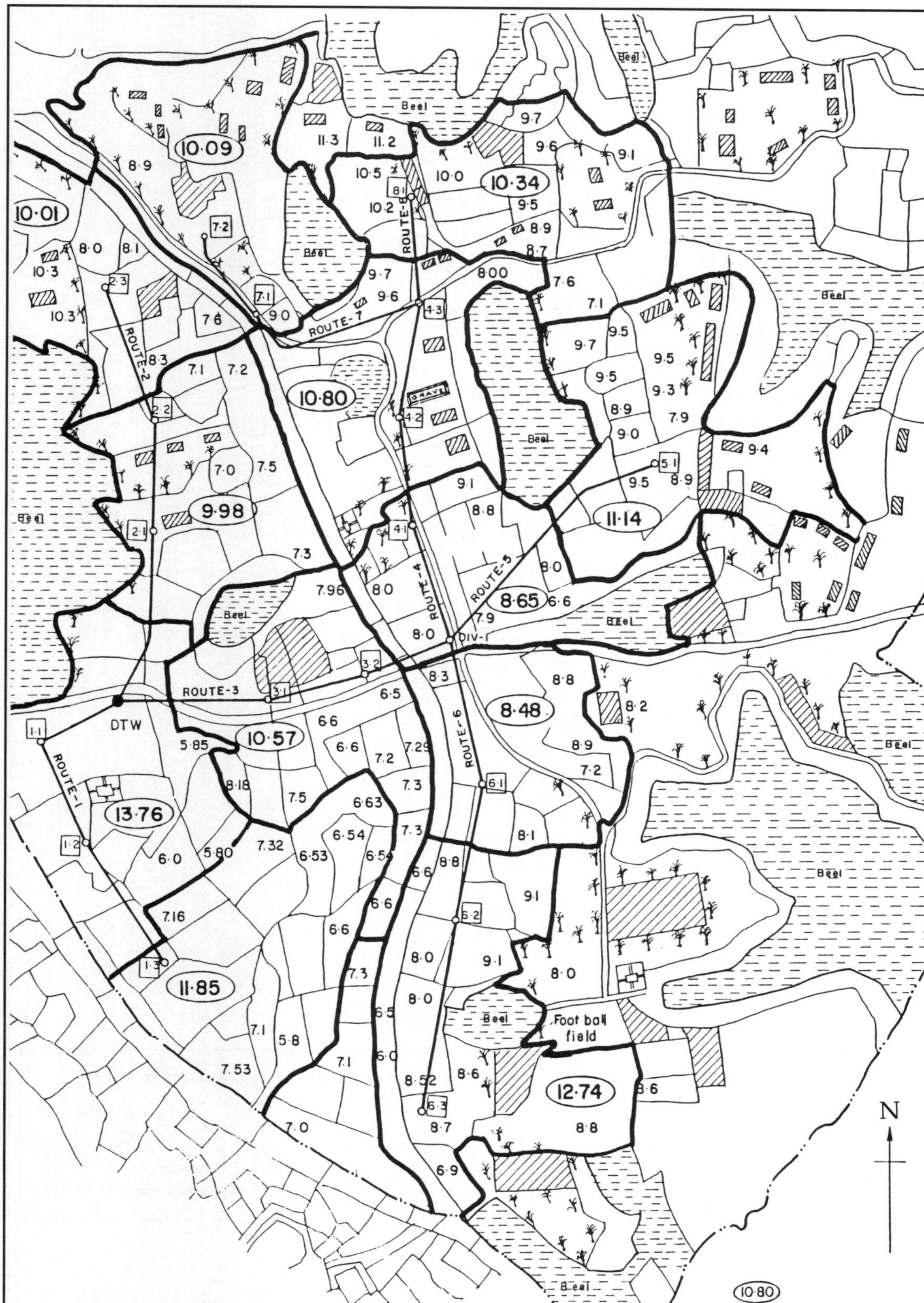

Total area = 129.02 acres

Figure 2.2. Branching pipe system layout, Bhaluka

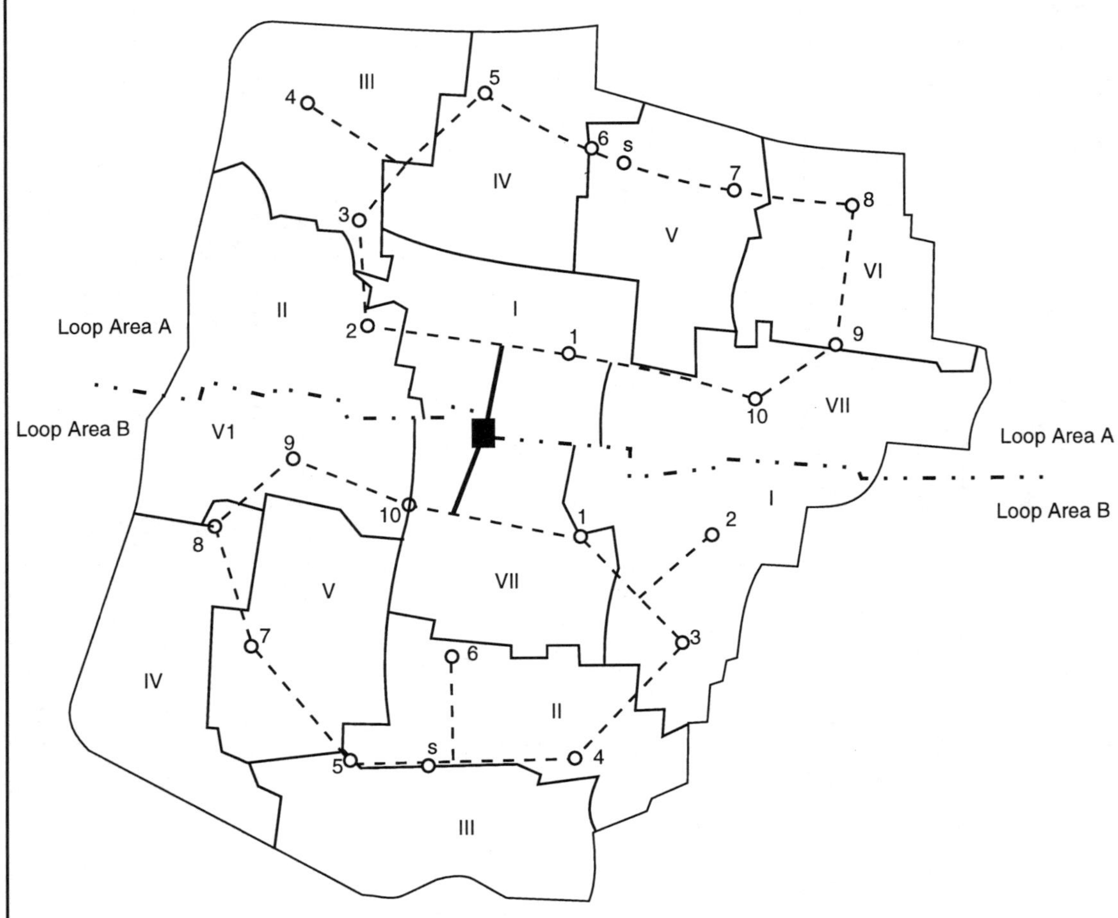

Area day commands	Symbol	Irrigated by outlet number	
		Loop A	Loop B
Monday	I	1.2.3	2.3
Tuesday	II	2.3	3.4.5
Wednesday	III	3.4.5	4.5
Thursday	IIV	3.5.6	7.8
Friday	V	1.6.7	5.6.7.9.10
Saturday	VI	7.8.9	9.10
Sunday	VII	1.9.10	1.6.10

■ Tubewell and distribution chamber
- - O - - Outlet valve
- - Ȯ - - Stand-pipe (s)
━━━ Feeder pipe
- - - - - Buried PVC pipe systems
——— Boundary of day area commands (about 7 ha)

Figure 2.3. Schematic layout for a uPVC loop pipe system

Source: World Bank (1983)

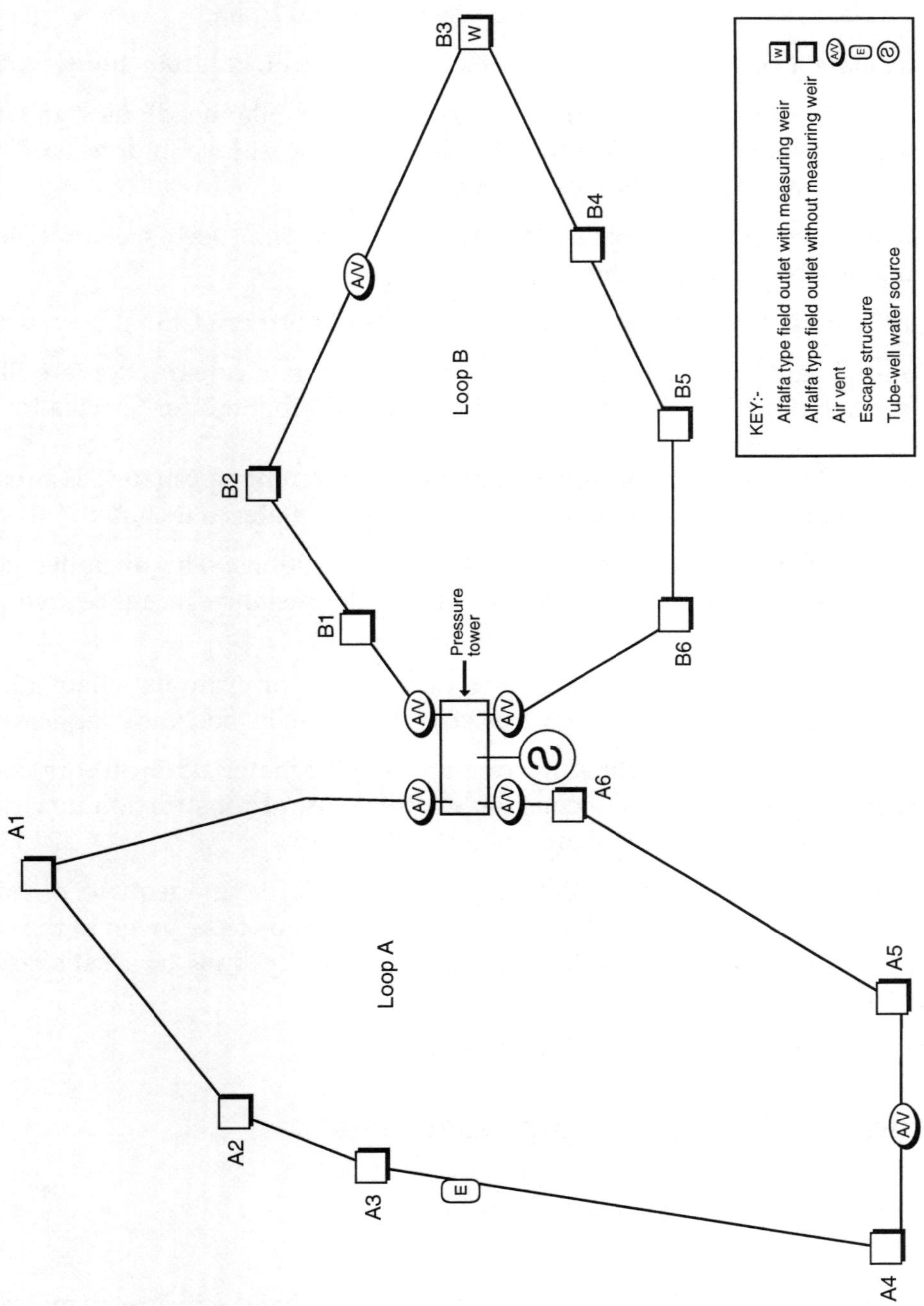

Figure 2.4. Typical buried pipe distribution system, plastic pipe

Source: Sir M. MacDonald and Partners (1989d)

Pipe material selection

Pipe materials which have been used for low-pressure pipe distribution systems include:

- non-reinforced concrete pipe; used in USA, Sri Lanka, China, India, Bangladesh and southern France;
- cast-in-place concrete pipe; used in China and USA;
- non-pressure class reinforced concrete pipe; used in central India;
- non-pressure class asbestos cement pipe; used in Bangladesh, Pakistan and Egypt;
- rigid uPVC pipe in pressure classes from 2.5 to 5 bar (25 to 50m head); used in Uttar Pradesh, India; Maharashtra and Madhya Pradesh, India; east Java, Indonesia; Thailand; southern Nepal; southern France; southern Spain;
- thin-walled corrugated smooth bore spirally wound uPVC pipe; used in central India, eastern China, Australia and southern Spain;
- thin-walled buried polythene hose; used in western and northern China.

Examples of different pipe materials used in buried pipeline construction are illustrated in the associated publication 'Photographs of Buried Pipelines for Surface Irrigation'.

Apart from the differences in transit efficiency between different pipe materials already described, important issues influencing selection of the pipe material include:

- Material suitability under the prevailing physical conditions. This includes pipe flexibility under soil settlement and movement, and resistance to aggressive soil conditions.
- For situations where seepage losses are to be minimized, for example where a loop layout is to be used, then uPVC pipe is often chosen despite its generally higher cost.
- The installed cost of the final pipeline. To compare the pipe materials on an equivalent basis, any cost comparison needs to consider the total cost of construction including transportation, installation, material and maintenance costs.
- The ease of pipeline construction. Where pipe materials have adequate physical properties and comparable installed cost, consideration needs to be given to the ease with which the particular pipe system can be successfully built to a satisfactory standard given the resources available.

Pipe material selection is summarized in Flow chart 2.8.

Physical suitability of various pipe materials

Pipe flexibility

General

Where pipelines are built on earth subgrade, and backfilled by hand with in-situ material, some settlement of the pipeline and backfill material must be expected.

Koluvek (1970) highlights problems in soils which are subject to cycles of movement due to changes in temperature and moisture, as this can lead to the disruption of joints, and on occasion pipes, in rigid pipe materials like non-reinforced concrete. This would apply particularly to organic soils and those with a high proportion of expansive clay, such as montmorillonite.

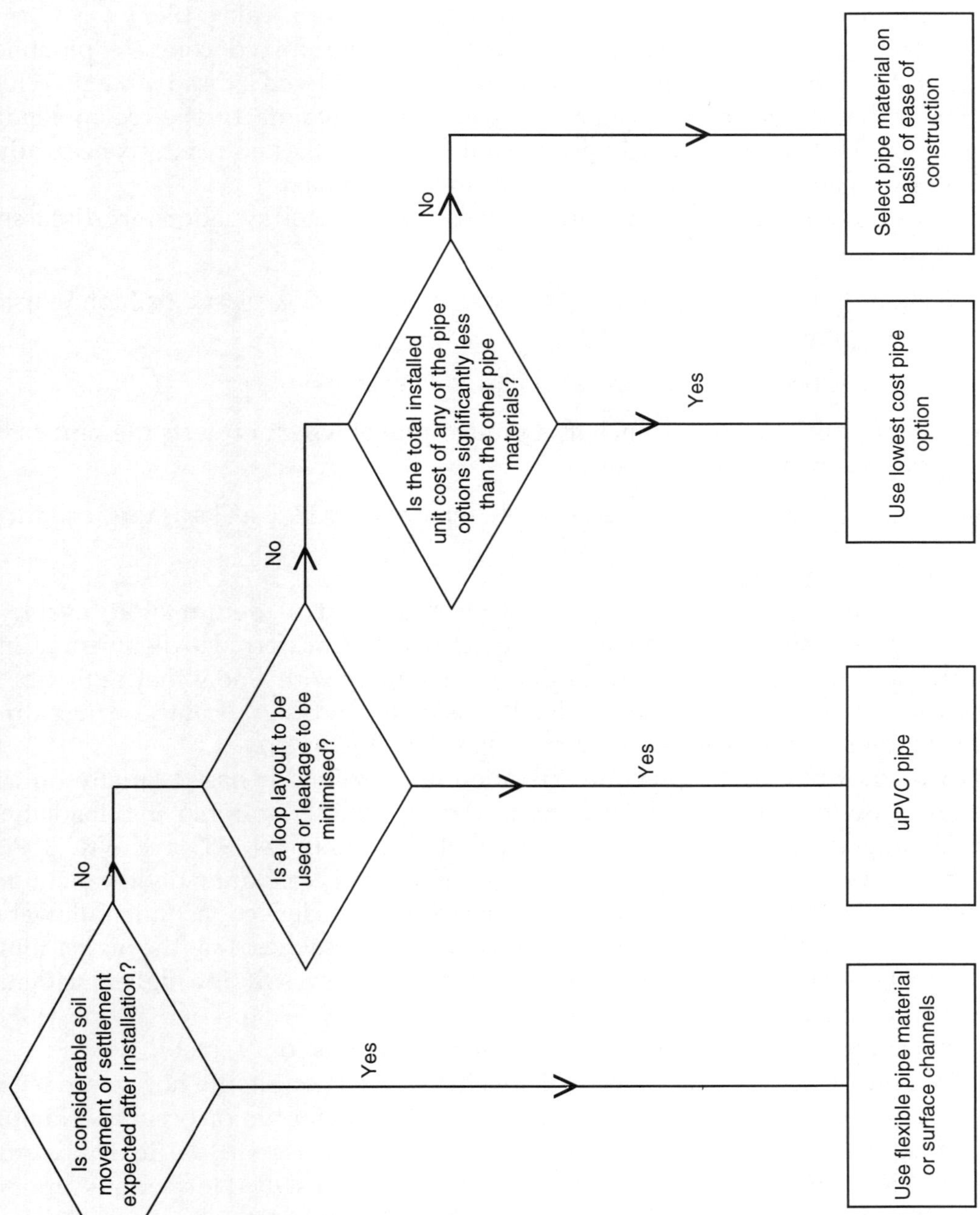

Flow chart 2.8. Pipe material selection

Options:
Polythene pipe
uPVC pipe
Surface channels

Note:
For concrete pipe in sulphate soils see Flow chart 2.9

Even in non-organic soils, work by Rashid et al (1992) provided evidence that significant numbers of leaks continued to occur after initial construction and testing, and it is possible that soil settlement contributed to the pattern of continuing pipeline leakage.

Another possible problem with concrete pipe in some areas is frost heave with shallow pipe embedment depths where very cold temperatures occur (Bealey, 1987).

Leak repair can be relatively inexpensive for non-reinforced concrete pipelines (Gisselquist, 1989), but it is inconvenient and must be considered a disadvantage. While uPVC pipe materials are certainly more flexible, current manufacturer's recommendations (Jain, 1989; HPPA, 1989) regarding application and installation are very conservative and do not as yet effectively address low-pressure applications.

Recommended strategies to help overcome these problem soil situations are discussed below and include:

- the selection of a more flexible pipe material such as uPVC plastic or high density polyethylene pipe in problem areas;
- ensuring pipe is bedded below any problem soil horizons;
- provision of backfill material, such as a graded gravel which protects the pipe from these soil movements;
- provision for the early diagnosis and timely repair of pipeline leaks as they occur.

Flexible plastic pipe materials

Flexible pipe materials, by allowing a controlled amount of deformation, enable a considerable proportion of the total external pipe load to be transferred to the surrounding soil. Plastic pipes are able to adjust to changes in soil-loading with time as they redistribute stresses to the adjacent soil. The degree of load transfer depends significantly on the nature and compaction of the backfill and surrounding soil (WRC, 1986).

Design for flexible plastic pipes in irrigation applications is based largely on the calculation of deformations or deflections of the pipe under soil and live loads and containing these within permitted limits (Järvenkylä 1988, Gumbel 1982, and WRC 1986). Although there is considerable agreement on the method of calculating deflection under load, there is also considerable divergence regarding the deflection limits allowable (Järvenkylä, 1988 and WRC, 1986). There is no evidence to suggest that the higher long-term deformation limits accepted in other European countries results in a higher incidence of pipe failure (Walton and Elzink, 1988). On the contrary Walton and Elzink (1988) suggest that current UK specifications include a very high factor of safety.

UK uPVC pipe manufacturers specify that vertical deflection should not exceed 6% in the long term (50 years), though as in practice most of the deformation occurs within the first two years, this can be used as the point in time at which to assess actual field measured deflection (WRC, 1986). Actual pipe deflection for pipelines installed for periods ranging from 14 to 20 years have been verified as being within allowable deflection limits (Walton and Elzink, 1988).

If design were modified to allow much greater in-situ deflection, then specifications for the installation and backfilling of thin-walled uPVC pipe materials could be relaxed considerably, with a possible reduction in the system cost. However, in the absence of agreement, manufacturers will continue to quote conservative design standards, for both short- and long-term deflections (Jain, 1991 and HPPA, 1989).

Pipe corrosion

Bealey (1987) concludes that aggressive soil conditions are rarely severe enough to result in durability problems for concrete pipe and he considers that it is really only of concern for concrete pipe materials, as plastic pipes are generally suitable for all irrigation water qualities normally encountered. Only particular waste waters when used for irrigation could present corrosion problems for concrete pipe.

Aggressive acidic soil or groundwater conditions seldom occur naturally and even then acid in contact with concrete pipe is neutralized and generally results in a neutral zone around the pipe (Bealey, 1987). Continued attack from acidic groundwater flow is only likely in some landfill sites and mining situations, when it may also present a problem for thin-walled plastic pipe materials (Jain, 1991 and Bealey, 1987).

The American Society of Agricultural Engineers (1989) highlights the problem of high sulphate soils in corroding concrete pipe. High sulphate salt concentrations are typical in some of the gypsum-rich soils of the dry western states of the USA and some parts of the Middle East.

Recommendations from ASAE S261.7 (1989) suggest that only where water-soluble sulphate levels are less than 0.1% should standard concrete pipe be used. In situations where water-soluble sulphate levels are between 0.2% and 1%, the use of special sulphate-resistant cement is suggested. For higher sulphate concentration soils, there appears little alternative to the use of an alternative pipe material to concrete (ASAE S261.7, 1989). The criteria and solutions recommended are summarized in Flow chart 2.9.

Loop layout or requirement for low seepage losses

Where a loop layout is chosen concrete pipe must be considered unsuitable. Reasons for advocating the use of branching pipe layouts for concrete pipe have been discussed above.

Loop systems which have been built have used either uPVC pipe (Cunningham, 1986; MMP, 1989a and IDTP, 1989) or asbestos cement pipe (WAPDA, 1989 and Plusje, 1981). Experience in India and Bangladesh suggests that uPVC pipe systems with very low seepage can be successfully built.

Installed cost of pipeline

When considering the cost of alternative pipe materials it is important to consider all factors which contribute to the cost of a unit length of installed pipe.

A full comparison should include:

- pipe material cost;
- pipe manufacturing cost;
- transportation cost including breakage;
- installation cost;
- pipe life where it can be estimated with some confidence;
- implications of different distribution efficiencies;
- differences in recurrent maintenance costs;
- implications of the proportion of imported and local material content.

Although the optimum choice of pipe material will depend on local availability and be very area-specific, a number of general recommendations can be made. The findings described below apply to pipe systems in south Asia and are based on cost information collected and presented in Chapter 8.

In all situations studied, uPVC pipe systems are considerably more expensive than non-reinforced concrete pipe, where both are available, over pipe diameters ranging from 150 to 300mm (MMP, 1989a and Gisselquist, 1986), even where the lower roughness of uPVC pipe is taken into account.

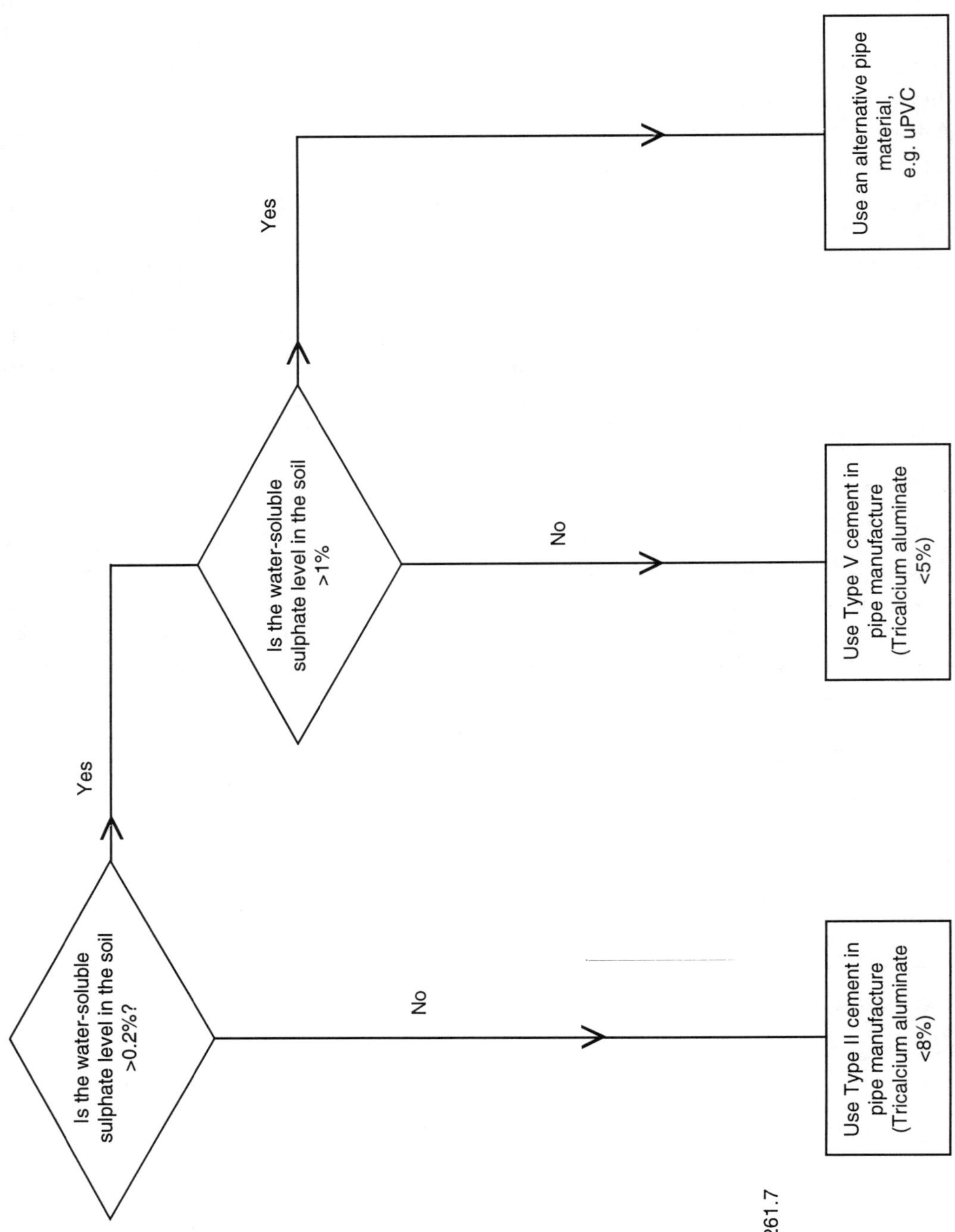

Source:
ASAE Standard S261.7
(1989)

Flow chart 2.9. Selection of concrete pipe in sulphate soils with water-soluble sulphate levels greater than 0.1%

More recent thin-walled uPVC pipe materials appear competitive with non-reinforced concrete pipe in sizes larger than 300mm, where transportation is a significant portion of the cost (Jain, 1989).

Non-reinforced concrete pipes will be competitive in price where local manufacture keeps transportation distances small, and an adequate quality of construction and jointing can be achieved.

Ease of construction

Consideration must be given to the ease of building a pipe system of adequate quality. The ranking of the low-pressure pipe materials, illustrated below, was developed based on an assessment of the ease of installation and using observations and information from local irrigation practitioners on the buried pipe systems evaluated in south Asia.

This takes into account the amount of supervision and quality control required to achieve a system with acceptable performance. Only pipe materials identified as being in use for low-pressure irrigation are included.

Easiest

rigid uPVC spigot socket cement-jointed pipe;

↓

thin-walled uPVC pipe materials;

↓

asbestos cement pipe;

↓

non-reinforced concrete pipe with tongue and groove joint;

↓

non-reinforced concrete pipe with plane ends;

↓

cast-in-place non-reinforced concrete pipe.

Most difficult

Where the cost and likely performance of various pipe material options is similar, then easier installation can lead to savings in the cost of supervision, reduced disruption to existing agricultural operations, and savings in the cost of repairing poor quality work at a later date.

Selection of system components

All buried pipe systems require a number of basic structures or components for the control and operation of the system. Structures are generally fewer in number than for equivalent open channel systems. In Chapter 3 specific procedures are outlined for the design of the pipe system and establishing the number, type and position of the structures required. Specific design guidelines for each component are referenced below.

Structures which are more elaborate than necessary increase the cost of the system and require more maintenance. In many cases a little modification is all that is necessary for a structure to assume several additional functions. Such modifications may include ensuring a minimum dimension is adhered to, as for example in the case of open stand-pipes in water hammer protection, or that an outlet distribution structure incorporates a v-notch or broad-crested weir for measurement of outlet discharge.

The selection of suitable structures is necessarily based on an understanding of their function. The structures required will be partly determined by the selections already made for regulating inflow discharge and for the operation and control of the system. Each of the common structures used in pipe systems is defined below and details are listed of the essential and additional functions performed by each structure. Refer to Chapter 3 for detail on design and examples of currently used structures.

Pumpstand and header tank

Stand-pipe structure open to the atmosphere connecting water source (typically tube-well pump discharge pipe) to the underground pipe network. The vessel extends from ground level to a height required to provide the necessary driving head in the pipe system. It is usually of a diameter sufficient to provide some limited storage buffer, significant reduction in water velocity and allow for fitting of valves on the inlets to the pipes.

Structures can be divided into two groups, firstly those where the water level is forced to rise above the pump discharge pipe and secondly those where the pump or water source discharges freely to the atmosphere.

Structures in the first group include pressure towers and head riser pipes, with low head pumpstands and elevated tanks included in the second.

Pressure tower

Pumpstand where water level is forced to rise to the design driving head inside the structure. Flow to the pipe system may be via the floor or through internal or external satellite stands. Structure diameter is usually sufficient to allow for the installation of valves or satellite stands inside the tower.

Essential

- Connect the water source to the buried pipe network.
- Provide the driving head for flow through the pipe network.
- Where there is more than one line, distribute the flow to different sections of the pipe system.
- Act as an air vent and surge riser for water hammer control.

Additional

- Provide for flow measurement for the entire or individual parts of the scheme.
- Provide for accurate flow division between two or more distinct parts of the command area.
- Allow for automation of pump operation using probes or float switches.
- Function as a sand trap.
- Provide some limited buffer storage to assist in inflow regulation.
- Allow for the fitting of valves on the inlets to the pipelines.

Head riser pipe

Small diameter vertical stand-pipe, between water source and underground pipe network, setting the maximum delivery head. It is equal to or larger in diameter than the underground pipe, acts as an air vent and surge riser, but provides no storage or room for the installation of a valve.

Essential

- Connect the water source to the buried pipe network.
- Provide the gravity head to drive the flow through the pipe network.

❍ Act to relieve pressure so that the rated pipe operating pressure is not exceeded.

Additional

❍ Provide for some limited buffer storage.

Low head pumpstand

Short open stand structure used to connect pump to pipe system where the driving head is at or just above ground level (up to about 1m). Pump discharges freely into the stand.

Essential

❍ Connect the water source to the buried pipe network.

❍ Where there is more than one pipeline, distribute the flow to different sections of the pipe system.

❍ Act as an air vent and surge riser for sudden valve closure on nearby outlets.

Additional

❍ Function as a sand trap.

❍ Provide for some limited buffer storage.

❍ Allow for the fitting of valves on the inlet to the individual pipelines.

A typical low head pumpstand is illustrated in Figure 5.1.

Elevated tank

Raised tank supplying one or more separate pipe networks via individual head riser pipes. Tank usually incorporates some mechanism for flow division and pump regulation can be automated where desired.

Essential

❍ Connect the water source to the buried pipe network.

❍ Provide the driving head for flow through the pipe network.

❍ Where there is more than one line, distribute the flow to different sections of the pipe system.

Additional

❍ Provide for flow measurement for the entire or individual parts of the scheme.

❍ Provide for accurate flow division between two or more distinct parts of the command area.

❍ Allow for automation of pump operation using probes or float switches.

❍ Function as a sand trap.

Stand-pipe structures

Overflow stand

An open stand structure connected to a pipeline in an open pipe system, at a position where excess head is to be dissipated. Water falls over a baffle to a lower water level on the downstream side. It may include a bypass gate in the baffle to reduce the quantity of air entrained during overflow.

Essential

- Dissipate excess head in open pipe systems by allowing water to fall over an internal baffle.

Additional

- Prevent pipelines which are higher in elevation than the pump discharge from emptying completely during periods when they are not operating.
- Provide for the controlled release of trapped air.
- Serve to reflect pressure waves during water hammer events.
- Function as a sand trap.

A typical overflow stand is illustrated in Figure 5.7.

Float valve stand

An open stand structure, on semi-closed pipe systems, housing a float valve which permits reduction in the pipe head or pressure while connecting, hydraulically, the upstream and downstream pipe sections.

Essential

- To provide a housing for the float valve which dissipates excess head in a semi-closed pipe system.
- To absorb flow momentum during changes in outlet flows on the pipeline.
- To prevent operational spillage when supply exceeds demand and allow demand to be sensed at the inlet to the pipe system.

Additional

- Provide for the controlled release of trapped air.
- Serve to reflect pressure waves during water hammer events.
- Enable automatic control of the water supply to the pipe system.

A typical float valve stand is illustrated in cross-section in Figure 5.9.

Gate stand

An open stand structure connected to the pipeline system whose function is to control the flow of water in some way. It may house two or more gates for controlling the flow of water to different pipe branches.

Essential

- To house gates to control the flow of water at a branch in the pipe network.

Additional

- To provide for the controlled release of trapped air.
- Serve to reflect pressure waves during water hammer events.

A typical gate stand is illustrated in Figure 5.13.

Surge riser

A vertical open stand to reflect pressure waves developed during water hammer. It is connected directly to the pipe system and rises to a height above the maximum level of the hydraulic grade line. It can also function as an air vent.

Essential

❍ Act as an open stand for the reflection of pressure waves developed during water hammer events.

Additional

❍ Allow for the controlled release of air which becomes trapped in the system without the loss of excessive amounts of water.

A typical surge riser constructed in Bangladesh is illustrated in Figure 5.16.

Air vents or valves

Open vents

Vertical open stand-pipe to allow for the release of air, connected to the underground pipe system and rising to a height above the hydraulic grade line for flow to the critical outlet. Used commonly where both the release of air and protection against water hammer are required.

Essential

❍ Allow for the controlled release of air which becomes trapped in the system without the loss of excessive amounts of water.

Additional

❍ Act as an open stand for the reflection of pressure waves developed during water hammer events.

Air valve

One-way valve allowing release of air but not water from the pipeline. Suitable for situations where open vents are not practical. Proprietary models available generally operate poorly at low pressure. Some models have been developed using plastic floats enclosed in a housing, which may perform better.

Essential

❍ Allow for the controlled release of air which becomes trapped in the system without the loss of excessive amounts of water where the use of an open vent is not practical.

Several types of air and vacuum release valves are illustrated in Figures 5.17, 5.18 and 5.19.

Riser

The section of pipe connecting the underground pipeline to the outlet valve. It is usually of the same diameter or smaller than the underground pipeline.

Essential

❍ Convey water from the buried pipeline to the outlet valve at the surface where it can be released if required.

Outlet valve

Adjustable exposed portion of the pipe system which when opened allows delivery of irrigation water. A number of different types of valves exist though the most common consists of a valve or plate mounted on a pipe riser and known as an alfalfa valve. Refer to Chapter 3 for detail on design and examples of currently used structures. General recommendations on the selection of outlet type are summarized in Flow chart 5.1.

Essential

- ❍ Deliver water from the buried pipe system at an adequate head for the field distribution system to be used.

Additional

- ❍ Allow the fitting of a portable hydrant for distribution of water by smooth or gated surface pipe.
- ❍ Allow the use of a flexible rubber hose for in-field irrigation distribution.

Various types of outlet are illustrated in Figures 5.22, 5.23, 5.24 and 5.25.

Hydrant

An above-ground attachment, either permanent or portable, fitting over an outlet valve (usually an alfalfa or orchard valve) which allows for additional control of irrigation water. Used as a means of connecting above-ground pipe or hose.

Essential

- ❍ Allow for the connection of the outlet to surface pipe or hose for in-field distribution.

Outlet distribution structure

The structure built around the outlet valve to control the discharge of water to the quaternary distribution system, which may comprise open channels or surface pipes or hoses.

Essential

- ❍ Allow the flow to be directed and distributed to one or more quaternary channels.
- ❍ Protect the surrounding area from scouring due to high flow velocities and uncontrolled discharge.

Additional

- ❍ Discourage physical tampering with the outlet, valve and riser.
- ❍ Provide for accurate flow division between one or more quaternary channels.
- ❍ Provide for the measurement of the outlet flow.

Two designs for outlet distribution structures as built in Bangladesh are illustrated in Figure 5.26 and 5.27.

CHAPTER 3

SYSTEM PLANNING AND DESIGN

Introduction

This chapter covers the technical procedures for planning and design of a buried pipe distribution system on a particular irrigation scheme. This stage follows the process described in Chapter 2 of selecting the following features, perhaps as the standard features (or options) of a project which covers many individual irrigation schemes:

- type of buried pipe system;
- the operating and control system to be used;
- the method of regulating the water supply which is envisaged;
- initial selection of the pipe material and layout to be used;
- the methods of quaternary distribution and irrigation to be used;
- the optimal outlet discharge.

However, during the design process any of these selections may need to be reviewed in the light of new information or decisions.

The chapter documents proposed design procedures and methods developed for low-pressure buried pipe systems as a result of the research work. The design of all pipe systems follows a similar procedure although elements such as hydraulic and structure design will differ for each type of pipe system. A worked example of a design for a closed pipe system distributing water from a deep tube-well supply is given in Appendix 2.

The overall planning and design procedure is summarized in Flow chart 3.1 and includes the following steps.

General scheme feasibility and pre-design survey

Feasibility study of the proposed scheme and pre-design survey will normally be undertaken as a separate exercise before the design process. This will include the general collection and analysis of physical, social and economic information relating to the proposed command area and analysis to ensure that the scheme is feasible in accordance with adopted criteria. Physical data could be based on an existing map base with checking and annotation to include relevant features. Social and economic data are usually collected by questionnaires or related techniques. Suitable procedures are discussed below.

Preparation of pipeline layout

A preliminary layout should be prepared following consultations with the farmers on the pipe layout, outlet positions and system operation. At this stage the aim of the layout would be to minimize the pipe length and achieve an optimum outlet density, without yet considering the hydraulic design. Pipeline layout design has been summarized into four steps detailed below, which are discussed in more detail later in the chapter and summarized in Flow chart 3.2.

- Establish the target irrigable command area.

- Set the maximum length of the quaternary channels.
- Establish the outlet service area.
- Select the simplest pipe alignment which satisfies design requirements.

Finalize system selection and obtain farmer agreement
The preliminary layout may need to be amended several times before it can be agreed, and the operating and control system and mode of operation finalized with the farmers.

Hydraulic design of the pipeline
As with all pipe design, pipes must be sized and designed to deliver the design flow within the allowable friction loss. With pipe distribution systems this is set by the difference between the head available at the inlet to the pipe system and the operating pressure required at the critical outlet. The head available for pipe friction will be determined by subtracting the minor losses for bends, valves and entry and exit losses at structures. Specific hydraulic design methods for open, closed and semi-closed systems are detailed in Chapter 4 and summarized in Flow charts 4.1, 4.2 and 4.3.

When a pipeline design and alignment have been decided, long sections for the pipe routes can be prepared.

Selection and design of remaining component structures
The pipeline design long sections are used to establish the need for structures for air release and water hammer protection and to detail their positions and dimensions. Details are given in Chapter 5.

General scheme feasibility

Before proceeding with the design and construction of a scheme it is important that all aspects of the development and its impact on the community are considered. The acronym FISHTEE can be used to represent the major aspects of development projects, as follows:

- Financial
- Institutional
- Social and cultural
- Health
- Technical
- Economic
- Environmental

All of these are important on irrigation schemes. In most of these respects, the feasibility study of irrigation schemes using buried pipe systems will follow the same procedures as for other irrigation projects. General procedures for planning and appraisal of irrigation schemes are outside the scope of this book, and will not be covered here. Useful references are Carruthers and Clark (1981) on irrigation planning and economic and financial appraisal, the Agricultural Compendium (Elsevier, 1989) on many of the common procedures used in feasibility studies, Coward (1980) on social aspects, Tiffen (1989) on health, Dupriez and DeLeener (1992) or Stern (1979) on technical aspects, Sargardoy et al (1982), Ostrom (1992) and Chambers (1988) on institutions, and Mock and Bolton (1991) on environmental assessment. Mention is made below of specific technical and socio-economic aspects which are important for buried pipe systems, by way of introduction.

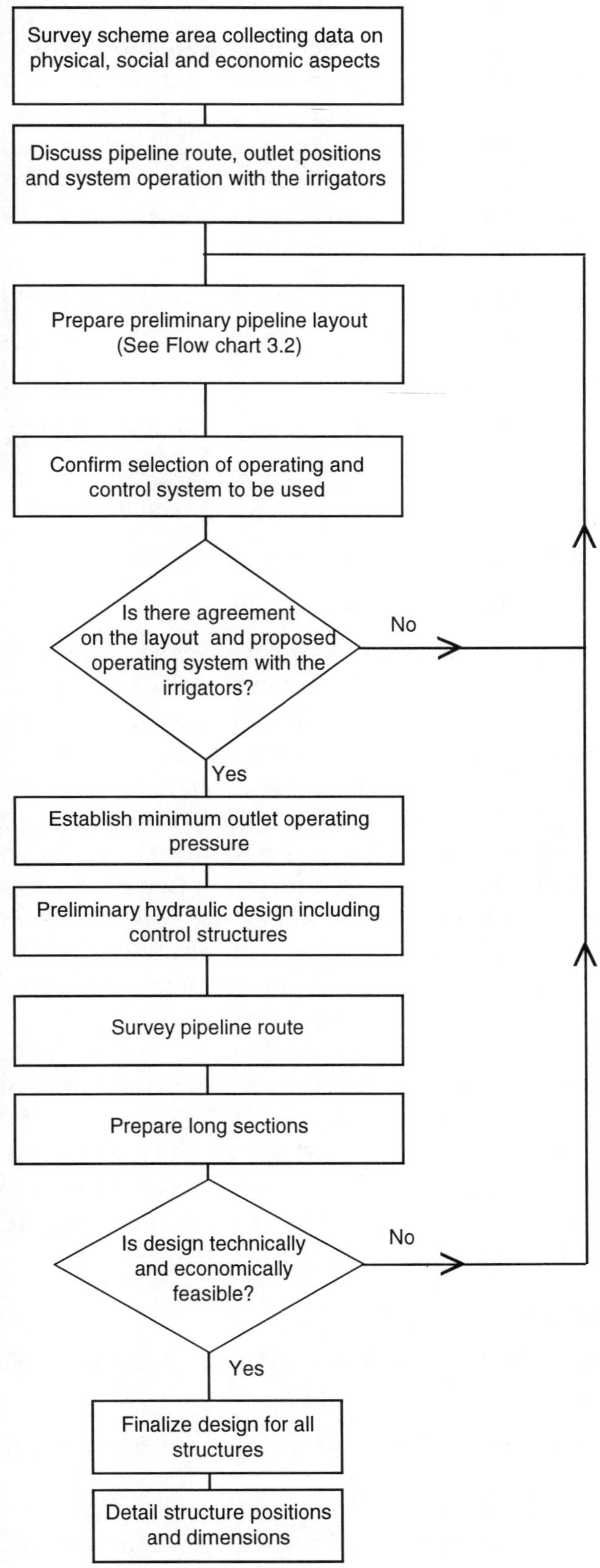

Flow chart 3.1. Design of buried pipe system

Survey information

Clearly the accurate design of a buried pipe system relies on having an accurate map of the potential command area with relevant plot boundaries, roads, tracks and other community infrastructure. Local land record maps are often a good first basis from which to begin though they will require updating through intensive field-checking and consultation with the irrigators. Other irrigation water sources in the area should be located to see how they may impact on the proposed command area.

Important features to be mapped include:

- land elevation as displayed by contour or relevant spot heights;
- boundaries for administrative areas, villages, individual land holdings and, where practical, individual plots;
- areas which will not be irrigated, such as homestead compounds, ponds, low-lying areas or forested land;
- roads, tracks and electricity or other services which cross the proposed command area;
- drainage features and other irrigation water sources;
- other relevant community infrastructure.

It is important that surveying is completed with the assistance of the farmers, and that they are given plenty of opportunity to comment on the design as it evolves. This requires effective communication and consultation procedures. Assisting the survey may be the farmers' first task on the scheme, and may begin the process which could lead to the formation of a representative water user association.

Buried pipe systems will require less intensive surveys than equivalent open channel systems. Knowledge of ground levels is required for the hydraulic design and it is normally sufficient to have spot levels for one or two points in each plot or field, including the highest spot to be irrigated. The extent and pattern of the existing field distribution system in the vicinity of each outlet should also be noted. It will be necessary for the designer to visit the field with the survey data and interpret these in the light of observation and in consultation with the irrigators.

The survey must be tied into a local datum, which could be referenced to a national datum, usually mean sea level. This may be particularly important in the interpretation of any water level information for tube-well-based systems.

Any survey should use a benchmark which will not be disturbed by any development which takes place. This may be the concrete floor of the pump house or a permanent dwelling or an alternative permanent feature. A permanent referenced benchmark will often be justified.

Social and economic assessment

Questions which could be asked when deciding on the development of any irrigation scheme include:

- Is water shortage the major constraint on the existing farming system?
- What are the other constraints?
- Is irrigation the most economical way of increasing agricultural production?
- Can existing schemes be improved more readily than a new project can be established?

- Given a limited supply of capital is an irrigation project preferable to some other form of investment?
- Is irrigated agriculture the most appropriate form of land use for the area?
- What are the operation and maintenance requirements of the scheme, and how will these be met — can the necessary funds and skills be mobilized?

If an irrigation scheme appears to be justified on these grounds, then consideration can be given to more specific social issues, including:

- seasonal demands for labour and its availability and distribution;
- effects on large and small farmers, tenants and labourers, and on the distribution of income and wealth;
- changes in the workload and status of women;
- changes in the distribution of responsibility and labour;
- impact on any inflationary pressure.

Social factors important in design

Apart from considering the overall impact of the development it is important to incorporate social concerns during the design of the system and to ensure that social conflicts are resolved. This is particularly important in the design of the layout, positioning of outlets, and determination of boundaries which will impact on land and water rights.

Land tenure and water rights

These are often interrelated, but land boundaries may not directly relate to water rights. There will often be some longstanding understandings between villages or households which can be easily broken by outside intervention. Frequently these land and water rights are not legislated. Customary rights are often different from statutory rights.

It is important to consult and to investigate thoroughly any existing uses for livestock (on land and water), fuelwood, downstream or upstream water users, and any burial grounds and sacred sites. In the semi-arid lands of Africa much large-scale irrigation development has taken place on prime grazing lands (Adams, 1992).

It is important to consider the difference between customary rights and ownership and whether government can appropriate land. It may be important to ensure that security of tenure is agreed to alleviate fears that government or others will take over land once it is improved.

Other important issues are women's rights to land, and whether farmer "failure" or non-cooperation should be punished with expulsion from the scheme and land.

Compensation should be considered for the loss of such items as:

- land through rights of way;
- houses, wells, trees;
- loss of crop during construction.

Whereas open channels require land aquisition for permanent rights of way along the route, with buried pipelines compensation is limited to any loss of crop during construction, which can be an important advantage.

Scheme layout

The layout should be discussed with both farmers and local officials, preferably on site. This needs to cover the proposed command area boundaries, pipe alignments and outlet positions, allowing the farmers to influence the decisions or choose between different technical options. The layout should be shown on the map, and on large-scale aerial photographs if available. If possible existing property and village boundaries should be respected as this eases management and eliminates tensions.

Sufficient time should be left in the planning and design stage to investigate boundaries and land rights, before any landmarks are destroyed. If farmers are allocated different plots after development every attempt should be made to retain the size and/or value of the original plot.

The planning issues include:

- ensuring the outlet and command area boundaries are consistent with community boundaries where possible;
- deciding strategies for allocating irrigation water in general and in times of shortage;
- maximizing the ability of the farmers to carry out the operation, management and maintenance of the system autonomously.

Any social problems existing at the time of the system's installation will probably continue to hamper the performance of the irrigation system. While inequity in distribution of water can be reduced by a buried pipe system, in practice this will rely on a collective will within the community.

Preparation of a pipeline layout

General

The following sections detail aspects which are important in designing an economical pipeline layout, following Flow chart 3.2. In preparing the layout the designer seeks to achieve a balance of large irrigated area, low capital cost, high distribution efficiency and convenient operation. These are all interrelated and dependent on other factors as well as the layout.

Pipeline layout design has been summarized into the following four steps which are discussed in turn below:

- establish the target irrigable command area;
- set the maximum length of the quaternary channels;
- establish the outlet service area;
- select the simplest pipe alignment which satisfies design requirements.

Target command area

The design command area of the distribution system may be calculated as follows:

$$A = \frac{Q}{q_d}$$

where A = design command area (ha)
Q = design inflow discharge to pipe system (l/s)
q_d = design irrigation duty (l/s/ha)

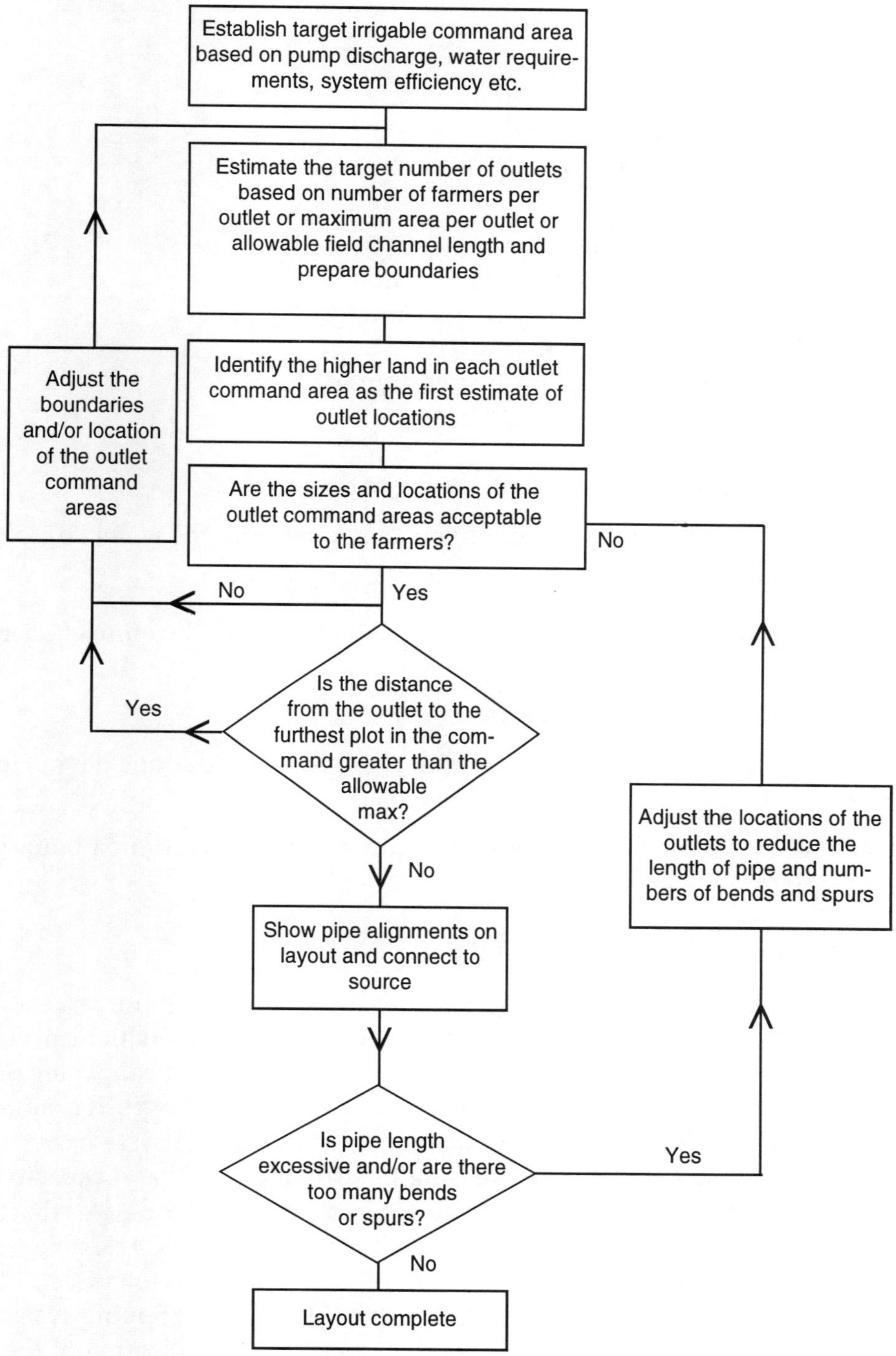

Flow chart 3.2. Preliminary pipeline layout

and $$q_d = \frac{I_f \times 10\,000}{U \times 24 \times 3600} \times \frac{1}{E_d}$$

where
I_f = field (gross) irrigation requirements (mm/d)
E_d = distribution efficiency (fraction)
U = utilization = duration of inflow as fraction of total time

also $$I_f = I_n \times \frac{1}{E_a}$$

and $$I_n = ET_{crop} - (P_e + G_e + W_b)$$

where
I_n = net irrigation requirements (mm/d)
E_a = application efficiency (fraction)
ET_{crop} = crop evapotranspiration (mm/d)
P_e = effective rainfall (mm/d)
G_e = gains from groundwater (mm/d)
W_b = gains from stored soil water (mm/d)

The size of the command area will therefore depend on:

- the design water duty and design inflow available from the water source;
- the cropping pattern and crop water requirements;
- the distribution efficiency of the pipe system and application efficiency of field irrigation;
- local climatic, soil and topographical features;
- local operating practices relating to hours of irrigation per day and days of irrigation per week;
- social and economic factors, which may limit the command area boundaries or the farmers to be included.

Available discharge and water duty

Setting the design irrigation duty (q_d) requires a decision about the peak water requirement which can be satisfied and hence the range of possible crops which can be grown. The design capacity of the pipe system must be sufficient for the peak water requirements envisaged, possibly with a safety margin to allow for future changes. It is more difficult to increase the capacity of a pipe system than of an open channel system.

There is also an argument for increasing the irrigation duty and capacity of the pipe system to increase the flexibility of operation. Merriam (1987c) considers that because one of the major advantages of pipe systems is their ability to provide a flexible water supply, limitations on the system capacity will not only be expensive to correct, but may also reduce the economic and financial returns to more intensive cropping systems.

For all systems a decision must be made regarding the utilization of the system (U) based on the practical duration of irrigation per day and per week particularly during the peak of the season. Operating practice ranges between large systems which depend on night-time irrigation to spread peak demand, and smaller systems where mainly daytime operation is practised. Although night irrigation allows the spreading of high demand peaks, this must be balanced against the lower level of efficiency achieved at the field level, and the unpopularity of working at this time.

For pumped supply systems the pump capacity and duration of its operation per day will set the peak water duty which can be satisfied for a particular command area. Although lower water duties with smaller pump installations or increases in command area can reduce investment costs per unit area, there is the risk of poor performance and conflict over water where supply is not adequate to satisfy peak demand within acceptable working hours.

Cropping pattern and crop water requirements

The design irrigation duty (q_d) reflects a choice of cropping pattern and corresponding water requirements (ET_{crop}). Field experience suggests buried pipe systems allow irrigators to manage more complex cropping patterns and cope more successfully with periods of water shortage than open channel systems, by allowing more precise flow division and rapid switching of flow from one part of the command area to another, with a minimum of wastage (Rashid et al 1992).

Pipe systems may also help to reduce average water duties by providing the flexibility to enable farmers to switch to more diversified cropping patterns (Campbell 1986, Gisselquist 1989), away from crops such as rice which consume higher levels of water.

Distribution efficiency of the buried pipe system

The replacement of open channels with buried pipe distribution systems in an existing irrigation scheme can be used to mitigate supply shortfalls or alternatively to increase the command area, as a result of the savings in water due to reduced transit losses. The overall distribution efficiency is the product of the transit efficiency of the tertiary (pipe) system and the efficiency of the quaternary system (often open channels) as discussed in Chapter 1.

Topographical features

Topographical features which can limit the command area of a pipe system include:

- the change in elevation between the source and the level of the highest field to be irrigated;
- the depth of a gully or drainage feature;
- a permanent body of water.

In practice there is seldom a problem except in severely undulating or dissected areas as it is often the feasibility of buried pipe systems in difficult topography which makes them preferable to open channel distribution systems. Where elevations or gully crossings result in the pressure rating of a pipe being exceeded, short sections of higher pressure rated pipe can be used.

Where command areas are sloping uphill, or are large in area, closed pumped-supply systems may be the only option. For pumped systems using pressure towers, such structures become both expensive and difficult to build successfully using local materials where their height exceeds 4 to 5m. In all cases care must be taken to ensure that the maximum design working pressure of the pipe material is not exceeded. This applies more particularly to non-reinforced concrete pipe than to uPVC pipe.

Where a pipe system is being installed on an existing pumped-supply system the additional head required to overcome pipe friction losses should not result in a significant decline in pump discharge.

Social and economic factors

In some cases technical considerations may be less important than social and economic factors which influence the selection of the command area. The importance of community participation in all irrigation schemes is generally acknowledged, but seems to be ignored too often by design engineers through the early stages of irrigation scheme planning.

It is necessary to establish local factors of importance, but some general recommendations can be made:

- Avoid excessively large command areas both at scheme and outlet level.

  These may create social tensions particularly where water supply does not match farmer expectation (van Bentum, 1992). Large scheme development may stretch local organizational structures beyond their capacity, particularly where structures are only weakly developed. Smaller pipe systems with fewer participating farmers are likely to have fewer organizational problems, and where the delivery stream is sized conveniently for the irrigator, design and operation is considerably simplified. Although capital costs per hectare will usually be higher for these smaller command areas, it is suggested that costs per hectare actually irrigated may prove to be lower because of the higher irrigation intensities achieved (Kolavalli and Shah, 1989).

- Do not cut across existing community boundaries, in terms of both their area and existing structures of organization and decision-making.

  Existing community structures will be important in providing organization and management of the system where no government-initiated structures exist or they are weakly developed.

- Proceed only when there is strong evidence of community commitment.

  Although evidence of community commitment may not be immediately forthcoming, any development should not proceed without strong support from a wide cross-section of farmers and landowners in the community. Consensus is needed, and situations of conflict between powerful members and factions in the community should be avoided.

- Involve irrigators at every possible stage of scheme selection and design.

  Involvement of the irrigators at every stage through planning, design and construction of the scheme will encourage a sense of ownership although it may slow scheme progress (Smout, 1990). This involvement can be structured as part of the formation of a water user association. Farmer responsibility for the operation and maintenance of the system may be fostered by commitment of a cash and/or labour contribution. Training of farmers to carry out any skilled work and conduct necessary maintenance procedures will also help foster co-operation. The process of being involved together in translating a pipe system from idea to reality can strengthen what may have been weak community links.

Maximum quaternary lengths

Almost all command areas where pipe systems might be used can be described by a regular shape approximating to a square, rectangle or circle. An analysis has been made of idealized pipe system layouts for these cases, and the results for branch systems are summarized in Table 3.1. This shows the pipe length per hectare for different quaternary lengths (l_4), command areas, and positions of the inlet to the pipe system.

Table 3.1 **Summary of pipe length and area data for idealized layouts**

Layout No.	Position inlet	Shape	Pipe length L(m)	l_4 = 50m Area A(ha)	L/A	Pipe length L(m)	l_4 = 100m Area A(ha)	L/A	Pipe length L(m)	l_4 = 150m Area A(ha)	L/A
1.	Centre	Circular	1160	13	92	2320	50	46.2	3480	113	30.8
2.	Centre	Circular	1060	13	84	2120	50	42.2	3180	113	28.1
3.	Rim	Circular	1405	13	112	2810	50	56.0	4215	113	37.3
4.	Rim	Circular	1235	13	98	2470	50	49.2	3705	113	32.8
5.	Centre	Circular	3075	28	109	6150	113	54.4	9225	254	36.3
6.	Centre	Circular	2585	28	92	5170	113	45.7	7755	254	30.5
7.	Centre	Square	1700	16	106	3400	64	53.1	5100	144	35.4
8.	Centre	Square	1100	9	122	2200	36	61.1	3300	81	40.7
9.	Rim	Square	1000	9	111	2000	36	55.5	3000	81	37.0
10.	Rim	Square	1700	16	106	3400	64	53.1	5100	144	35.4
11.	Centre	Rectangular	900	8	113	1800	32	56.3	2700	72	37.5
12.	Centre	Rectangular	700	8	88	1400	32	43.8	2100	72	29.2
13.	Centre	Rectangular	2250	18	125	4500	72	62.5	6750	162	41.6
14.	Centre	Rectangular	1850	18	103	3700	72	51.4	5550	162	34.2
15.	Rim	Rectangular	950	8	119	1900	32	59.4	2850	72	39.6
16.	Rim	Rectangular	800	8	100	1600	32	50.0	2400	72	33.3
17.	Rim	Rectangular	1950	18	108	3900	72	54.2	5850	162	36.1
18.	Rim	Rectangular	1750	18	97	3500	72	48.6	5250	162	32.4

Note: l_4 *= quaternary length*

Optimizing the pipe layout will involve achieving a desired outlet service area and density with a minimum length of pipe per hectare of command area (L/A). Table 3.1 shows that L/A appears to be inversely proportional to the maximum distance from any outlet to the furthest field commanded by that outlet. This straight-line distance is described as the maximum quaternary length (maximum l_4), though in practice quaternary field channels will be longer as they bend around field boundaries.

Therefore, as the maximum quaternary length (maximum l_4) increases, pipeline length per hectare (L/A) decreases. This increased quaternary length usually reflects a decrease in the number of outlets, and hence an increase in the outlet service area, so that a shorter pipe network is required to link fewer outlets. This needs to be kept within the desired outlet service area (see below).

Increased quaternary length also implies increased quaternary losses, and therefore a lower overall distribution efficiency (E_d). The designer needs to select a quaternary length based on a target distribution efficiency, management and operational considerations and estimated losses per unit length of quaternary. These losses are influenced by:

- the type of quaternary system (typically earth channels but surface pipe and hoses are also used);
- the soil type;
- the quality of construction of the quaternary system (permanent or rebuilt each year).

These are discussed in turn below.

Type of quaternary distribution system

The choice of the quaternary distribution system will depend on the crop grown, plot size and irrigation method used. Where large plots and good land grading allow graded furrows and borders to be used, distribution channels or surface pipelines need only be short, compared to the more extensive channel networks required to supply highly fragmented basin irrigated plots.

Where surface pipes, gated pipes or hoses are used, the distance between outlet and plot will be limited by the cost of the pipe or hose material rather than the quaternary distribution efficiency. This cost is reduced where a number of irrigators can share the use of the surface equipment.

Earthen channels, though the least efficient of the distribution alternatives, generally have the lowest capital cost. They should only be used over short lengths (usually not more than 100m) except where there are low-permeability soils and where there is good co-operation among farmers and a willingness to carry out timely and adequate maintenance.

Soil type

Earth channels constructed from poorly cohesive and coarse-textured soils tend to have lower distribution efficiencies. Even for fine-textured silt loams in Bangladesh, measurements of earthen field channel distribution efficiencies indicate approximate transit losses in the order of 5 to 8 l/s/100m of channel with flows of 40 to 50 l/s (Rashid et al, 1990 and 1991; MMI, 1990b). This is some 5 to 10 times the loss rates on the non-reinforced concrete buried pipe systems (Rashid et al, 1992). Any gains from an improvement in transit efficiency by the use of buried pipes can be lost with excessively long earthen field channels. Based on the Bangladesh data, even with low infiltration rate soils the field channel lengths should not exceed 200m, and they need to be substantially less on coarse textured soils.

Construction quality of the quaternary system

The degree of compaction of the banks of an earth channel has a strong influence on the seepage losses from it. Adequate sizing and bank top levels are also important to prevent overtopping, and care of outlet points is needed to prevent leakage. These factors are related to the permanence of the channel and its maintenance.

Field observations in Bangladesh and India indicate that the crops and cropping pattern strongly influence the quality of the earthen field channels. Where widespread basin irrigation of rice is carried out, field channels are commonly rebuilt on an annual basis with very limited compaction (Rashid et al, 1992). Where a wider range of crops is grown throughout the year, the field channels tend to be a more permanent feature.

Early assessment of the likely performance of the in-field distribution system at the design stage will help in establishing a realistic quaternary length.

Outlet intensity

Factors influencing the service area per outlet include:

- the acceptable quaternary length;
- the number of landholders and the extent of land fragmentation;
- the proportion of non-irrigable land close to the outlet.

These are discussed in turn below.

Quaternary length

Although with increasing maximum length of the quaternaries the service area of each outlet rises, actual service areas vary considerably. For maximum quaternary lengths (l_4) of about 200m, the outlet service area is normally in the range 2 to 4 ha. The service area falls below 2 ha only in situations where l_4 approaches 100m. The outlet service areas of uPVC pipe systems which have been constructed are larger than comparable non-reinforced concrete pipe systems and this reflects quaternary lengths which are often in excess of 250m.

Number of landholders

If the number of landholders in the vicinity of an outlet is large, often because of significant fragmentation, operation of the system may be simplified by providing an additional outlet. This can also help to reduce the extent and complexity of the field channel network. In command areas where landholdings are highly fragmented the field channels often take tortuous routes and are narrow with sharp bends. Land consolidation is frequently difficult to implement but is a way to enhance the benefits of a piped system of water distribution by rationalizing the earth channel network, and considerably simplifying operation of the system. Conversely, highly fragmented holdings will often justify a smaller outlet service area than would be suggested for technical reasons.

Distribution of irrigable land in the service area

In order to maintain realistic quaternary lengths, schemes where irrigable command is fragmented will require more outlets, with smaller service areas, than suggested for consolidated service areas.

Pipe alignment

The analysis of idealized layouts showed that the pipe length per hectare is influenced most strongly by the quaternary length and outlet service area, and only small differences result from variations in shape of the command.

General findings

Whether a loop or branch-type layout is used, the length of pipe is largely determined by the maximum acceptable distance from any outlet to the furthest field commanded by that outlet (l_4).

For both loop and branch-type layouts, as the maximum quaternary length increases then pipeline length per hectare decreases. This usually reflects a decrease in the number of outlets, and hence increase in the outlet service area, as a shorter pipe network is required to link fewer outlets.

The factors important in determining the pipe layout include:

1. The position of the inlet or water source in relation to the command area. This in particular relates to whether the water source is on the rim or near the centre of the command area.

   For circular or irregular-shaped commands, given the same quaternary length, a water source at the perimeter increases the pipe length by up to 20% compared with a more central position. For more elongated-shaped commands the change in pipe length will be site-specific.

Other factors which will influence selection of the position of the water source in relation to the command area include:

- distance from power supply for electric motor-driven pump supplies: It may be cheaper to bring the pump to the power source rather than the reverse, even when the implications on the cost of the distribution system are considered;
- distance from village settlement for ease of supervision and prevention of vandalism;
- location of high land to facilitate gravity supply or lower elevated tank heights.

2. The proportion of the gross command area which is not irrigable. A large non-irrigable area will result in a larger gross command area, requiring a more extensive pipe network to achieve the design net command area.

   In such cases, particularly if the irrigable land is fragmented, it may not be economically sensible to invest in a single tertiary distribution system for the whole area. An alternative may be to provide smaller systems in separate parts of the command area with higher proportions of irrigable land. An example of this would be the use of a larger number of smaller capacity tube-wells, as an alternative to the deep tube-wells widely installed in south Asia.

3. The discharge of the system and therefore the target command area. Increasing the size of the command tends to increase the pipe length required per irrigated hectare as the average distance of the outlet from the water source increases.

   Conversely, larger command areas allow the cost of the water source and control structures to be spread, resulting in lower costs per unit area. The cost decrease outweighs any increase due to the extra length of pipe per hectare. This can be an advantage in systems utilizing a deep tube-well or where the extent of an expensive primary and secondary network can be reduced.

Method of pump regulation and system operation

The methods of possible pump regulation on pipe systems are outlined below. Final selection should take into account the preferences of the irrigators, their level of skill and the opportunities provided by a more automated level of control. The options are divided into manual and automatic systems.

Manual pump regulation

Labye et al (1988) define and describe two methods of manual regulation involving either simply co-ordinating the operation of the system (simple method), or using a balancing tank or pressure riser (water level observation).

Simple method

The operator starts the pump once the outlet requiring water is open and then adjusts the engine speed or outlet opening to achieve a balance. Overflow of a control structure such as pressure tower, tank or air vent will indicate the need to reduce the pump inflow discharge or increase the outlet discharge. There is no storage in the system to help buffer variations in demand against supply. Such a system will need to incorporate a pressure release mechanism when the line operating pressure exceeds the recommended pipe working pressure; for example, if the pump is started with all the outlets closed (Labye et al, 1988). This could include:

- provision of head riser pipe between the pump and the pipe network which would overflow in the event of excessive line pressures (this has been used in the pipe systems built in China and Indonesia as reported by DST, 1990 and Republic of Indonesia, 1990);
- a pressure-release valve triggered for a particular line pressure (this was used on directly connected pipe systems in Thailand as described by HHP/HTS, 1988); more details of these types of valves are given later in the chapter;
- an in-line pressure sensor which would switch the pump off at a pre-determined trigger pressure.

As the situations mentioned illustrate, the simple method is used commonly where the pump is connected directly to the pipe network and one irrigator usually takes the full pump discharge. Close matching of the pump selected with the required pump duty in the system helps to simplify regulation. The most suitable pumps will be those with relatively flat pump curves as they will satisfy varying outlet discharges without developing excessive line pressures.

Water level observation

The distribution system utilizes a pressure tower or riser pipe between the pump discharge pipe and the pipe distribution system, which provides some amount of buffering storage and allows the pump to be regulated within a desired head range or water level. A flagged float or sight tube may be provided at the pressure tower or head riser pipe to assist the operator in regulating the water level.

In some cases a larger capacity tank is provided and the operator controls pump operation to ensure the tank is filled, so providing for irrigation over an extended period. If the tank is distant from the pump it is usually equipped with a float or water level indicator which can be seen from the pumping station.

Automatic pump regulation

Automatic pump regulation is only an option for electric motor-driven pumps and must include pressure and electrical sensing equipment to protect the pump and pipeline.

Labye et al (1988) define and describe three methods of automatic pump regulation:

- using an elevated open tank or reservoir;
- by air pressure vessel;
- by flowmeter associated with an air pressure vessel.

Regulation using an elevated open tank

This is the most widely used method in buried pipe distribution systems. Two types of control are commonly used:

- pressure sensor regulation, and
- water level regulation.

Although in most cases the tank is located near the pump it may also be located at the other end of the network where the site is suitable.

Pressure sensor regulation

This system requires an elevated balancing tank, a pressure sensor mounted on the pump delivery pipe, a pressure controller which stops the pump at a preset pressure, and a time delay switch which restarts the pump after a preset time period. A float valve on the discharge pipe entering the tank closes as the water level rises and causes the pressure switch to stop the pump. After the time delay the pump restarts if the line pressure is lower than the pressure switch value.

Problems could arise with the failure of the float valve to close adequately. The tank needs to be large in comparison to the pump delivery flow rate (holding three to five minutes flow) if rapid pump cycling and wear of components are to be avoided.

This type of control is relatively inexpensive and simple but only suited to single pump installations on small irrigation schemes, with flows less than 10 l/s. This system has not been recorded in use on any low pressure buried pipe systems.

Water level method

This involves an elevated tank fitted with float switches or probes which are connected to the pumping station by line or radio. This type of control has been provided on the uPVC pipe systems built in Uttar Pradesh, India (Campbell, 1984 and IDTP, 1989). The frequency with which the pump is stopped and started is dependent on the tank volume and the variation in demand. Very frequent stopping and starting is detrimental to the electrical and hydraulic equipment. Careful sizing of the tank can avoid most problems, though a time delay switch can also be fitted as a final control on pump operation.

This system is really only feasible where the regulating tank is close to the pump and water source to enable electric cable connection to be used. Radio control is possible but very expensive and not as reliable.

Because the tank must be elevated, this method becomes a relatively expensive option in the absence of high ground adjacent to the pump. This is likely to make this option too expensive for small irrigated areas (under 40 hectares) given the high cost of building the superstructure for the tank.

Air pressure vessel method

This method is applicable to small pipe systems operating at medium and high pressure. In this case the pressure vessel replaces the elevated tank. The pump is set to stop and start at preset pressures as sensed by the pressure switch. Since the cost of the air pressure vessel is a function of volume and is usually high, the size of the vessel should be kept as small as possible.

For low-pressure pipe systems for irrigation this type of automation is not practical because flow control depends on the storage provided by an adequate range in operating pressure in the pressure vessel. Although simple and of low capital cost when the pump discharge is small, it is not suited to surface irrigation distribution systems with their larger flow rates. Because the system requires that the pressure vessel operate over a pressure range above the nominal system pressure, it may also necessitate the use of higher cost pipes and fittings, so increasing the capital cost.

Regulation by flowmeter

This method utilizes:

- an electrical flowmeter mounted on the main pipe;
- a small base pump which is controlled by a small pressure vessel and provides a low discharge to make up for losses in the system;
- a number of pumps with flat head-discharge curves and capable of satisfying demand over a wide range of flows.

The pumps are stopped and started according to demand as measured by the flowmeter. A balancing tank is not required which is an advantage where the tank would have to be elevated. If the pumps are carefully selected the pumping head will be lower than that required for regulation using an air vessel. However, very accurate flowmeters designed for these conditions are expensive. In general a multiple pump system will have manual operator control as a backup.

This type of system could be practical on larger buried pipe systems with a surface water source. A buried pipe system utilizing multiple pumps with manual control has been built in Egypt and is described by EWUP (1983).

CHAPTER 4

HYDRAULIC DESIGN

Introduction

For the purposes of hydraulic design each distinct part of the pipe network is considered separately.

For open and semi-closed pipe systems each pipeline section between the overflow or float valve stands must be designed individually. However, where ground slope and operating pressures are similar, pipe design for different pipe sections will be equivalent. Closed pipe systems are considered as one unit, though different branches or loops from the same water source are analysed separately.

When designing the pipe system and calculating the hydraulic gradient the following points should be remembered for all systems (Koluvek, 1970):

- The hydraulic gradient should not drop below the pipe elevation at any point where water is flowing.
- Design should be based on operation of the critical outlet, which is usually the outlet most distant and/or highest in elevation relative to the water source.
- For branching systems, design should begin at the downstream end of each branch, and for sections above junctions with stand-pipes, design should be based on the highest gradient required.

Note: The hydraulic gradient is an imaginary line representing the height that water would rise in an open column, at different points along the pipeline. The line begins from the height of water or the driving pressure at the start of each separate pipe network and the slope will vary with the velocity of flow in the pipeline. A typical hydraulic gradeline is illustrated in Figure 4.1.

Design velocities

Maximum velocities

The choice of velocity results from a compromise between a high velocity to reduce the diameter and cost of the pipe and the higher cost of water hammer protection associated with higher velocities.

Recommended maximum velocities for flow in low-pressure pipelines are in the range of 1.3 to 1.5m/s (ASAE S261.7, 1989). This applies to the mortar-jointed non-reinforced concrete pipe, low-pressure asbestos cement, and thin-walled uPVC materials.

For most closed low-pressure pipe systems, the use of low-pressure pipe materials (with maximum operating pressures < 10m) and open stand-pipes for control structures will mean that pipe velocities seldom exceed the maximum recommended velocity of 1.5m/s. For the buried pipe systems in India and Bangladesh with driving heads in the vicinity of 3 to 5m, typical flow velocities range between 0.6 to 1m/s. Velocities of flow in excess of this range result in much greater pressure rises due to water hammer, requiring more frequent and costly provision of surge risers.

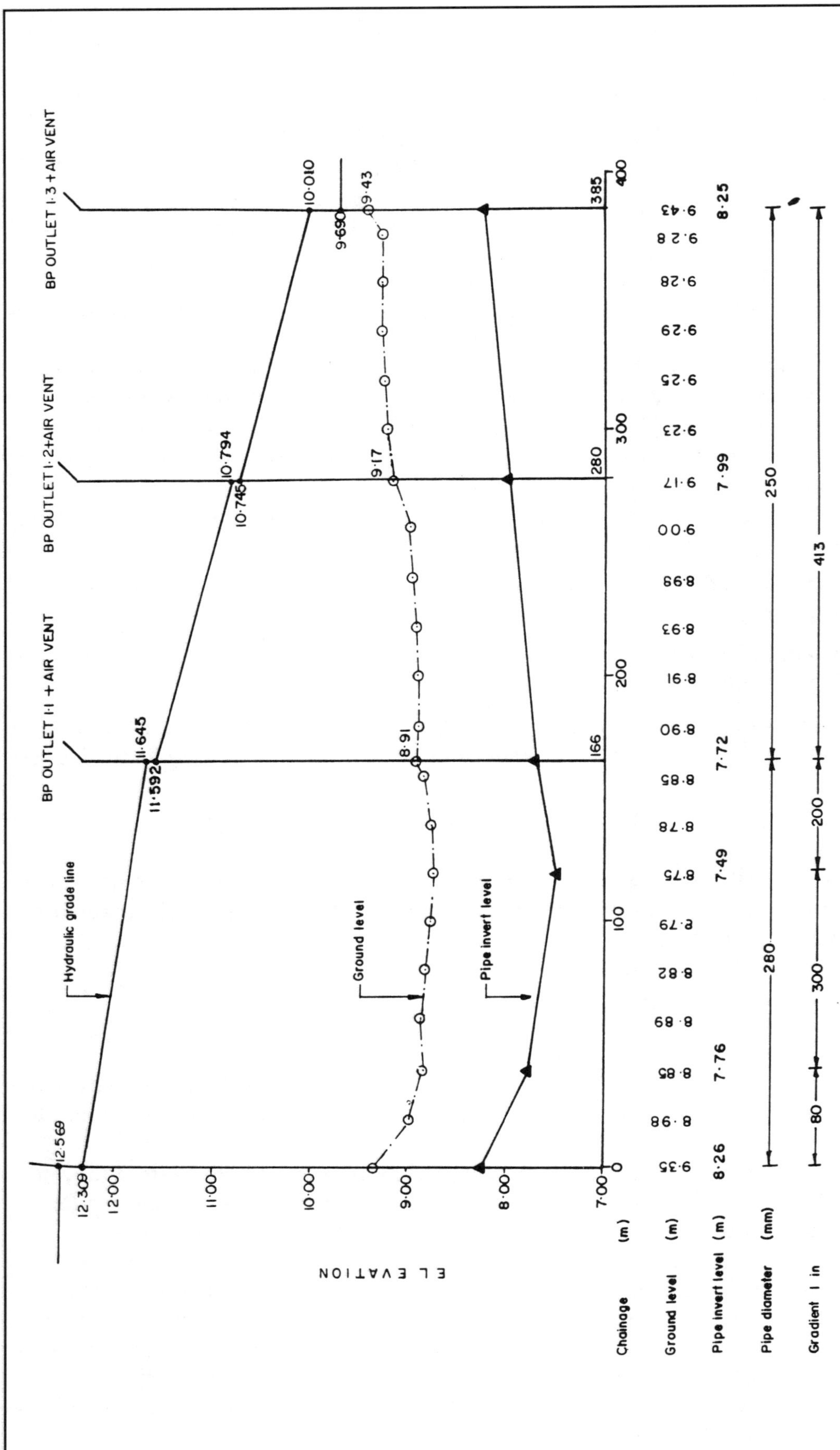

Figure 4.1. Pipeline long section and hydraulic grade line

Table 4.1 **Minimum permissible velocity in pipes**

Diameter (mm)	*Min. velocity (m/s)*
350	0.50
300	0.40
250	0.40
200	0.35
150	0.25
125	0.25
100	0.20

Source: Labye et al, 1988

Where the pipe materials used allow higher operating pressures (> 20m), higher flow velocities are possible. This is possible with plain-walled uPVC pipe and reinforced concrete pipes with rubber ring joints where maximum flow velocities can be as high as 3m/s. These pipe systems are effectively no longer low-pressure systems and careful consideration must be given to water hammer and the adequate venting of air.

In the case of open and semi-closed pipe systems where slopes are in excess of those required to overcome pipe friction losses, it is essential to select the pipe size so that the recommended pipe velocity (1.5m/s) is not exceeded. Head in excess of the pipe friction loss at design flow is then dissipated in the respective overflow and float valve stands.

Minimum pipeline velocities

Recommended minimum velocities are based on the need to ensure that any sediment or debris entering the pipe system is flushed during normal operation. Under particular operating conditions flow velocities will drop below these guidelines but material should in most cases be scoured adequately when normal flow regimes resume.

Standard texts indicate scouring velocities for non-cohesive materials in the range of 0.3m/s for silts and 0.5m/s for fine sands, which are likely to be the most troublesome materials entering pipelines.

Labye et al (1988) state that for pressure pipe networks, the minimum permitted pipe velocity depends to some degree on the diameter of the pipe. A velocity of 0.5m/s is acceptable for pipes of 350mm diameter or greater, but lower velocities are permitted for smaller pipe sizes as outlined in Table 4.1. However, wherever significant sedimentation is expected, a sand trap should be provided, otherwise design flow rates should not be less than about 0.5m/s.

Pipeline slopes

Low-pressure pipelines will not be practical in very steeply rising or falling command areas, because of both the operating pressure limitation imposed by the low-pressure pipe materials and the high cost of providing frequent water pressure control structures. Low-pressure systems will be practical over the whole range of land slopes from 0 to 10%, though higher-pressure pipe systems are likely to become competitive at the steeper slopes.

Short sections of quite steep pipelines can be installed provided adequate provision is made for any axial thrust forces. Extra pressure control and air venting structures may also be required. Semi-closed and open pipe systems will usually be laid at positive slopes, with adequate opportunity for air movement.

Closed pipe systems on very flat topography can present problems with air movement. Reverse slopes on pipelines are acceptable, though repeated slope reversal should be avoided, as it creates high spots where air will tend to accumulate and need to be vented.

Conversely laying pipelines nearly flat should be avoided where possible as it tends to increase any surge problems associated with trapped air. Where this cannot be avoided more frequent air vents or large orifice air valves may be needed.

Twort et al (1985) suggest a minimum slope (in the direction of flow) of 1 in 300 for a descending pipe slope, and 1 in 500 for a rising pipe slope, to ensure that air moves satisfactorily to any air vents provided. However, they acknowledge that this is often an impractical ideal where pipelines are installed in level terrain.

Hydraulic design of open and semi-closed pipe systems

Before pipeline design is possible, on open and semi-closed systems, it is necessary to determine the overflow stand height and spacing on open systems and the float valve stand height on semi-closed systems. Pipe size and hence pipe friction losses must be determined for each section of pipeline between stands. However, as both overflow stands and float valve stands are designed to dissipate head in excess of rated pipe pressure, friction losses for the pipe sections will clearly be less than the change in ground elevation for the pipeline section upstream of the structure.

Determining the spacing and pipe size for open and semi-closed systems
This can be summarized in two steps:

Step 1. The spacing between stands is given from the equation

$$M = E + D + s_g L$$

where M is the maximum safe pressure for the pipe
E is the minimum energy head required at the stand
s_g is the ground slope
L is the stand spacing
D is the drop in water level at each stand

For open pipe systems the maximum stand height must also not be exceeded. For detail on the design of both open stands and float valve stands, refer to Chapter 5.

Step 2. Pipeline selection
Given the spacing between the stands, an appropriate size of pipe must be selected such that the flow velocity is less than 1.5m/s. The head to be dissipated at the float valve can then be calculated.

The design procedures differ for each type of system and are explained in detail in later sections.

Summaries of the hydraulic design procedures are presented for open systems in Flow chart 4.1 and for semi-closed systems in Flow chart 4.2.

Hydraulic design of closed pipe systems

The hydraulic design of closed pipe systems relies on identifying the critical path for each distinct pipe network. This will often mean analysing the head losses for the design flow to a number of outlets (usually on or near the perimeter of the network). Minor losses and the minimum operating pressure at the critical outlet are subtracted from the available upstream head to determine the head available for pipe friction losses. The design is then completed for all the remaining parts of the network. Each of the elements in the hydraulic design is discussed in more detail in the sections which follow.

Flow chart 4.3 provides a summary of the procedure for hydraulic design for closed pipe systems. A worked example of system design for a closed pipe system is described in Appendix 2.

Available upstream head

The driving head available will be determined by the maximum height of the controlling water level at the water source.

For pumped supply systems this will be set by the maximum practical height of the control tank or pressure tower, the pressure rating of the pipe and the head/discharge characteristic for the pump. The limit to the height of the pressure tower or riser pipe will often be a trade-off between increasing cost of the structure with height and reduced discharge and/or increased pumping costs with height.

On gravity-supply systems available head will relate to the expected lowest operational water level from the water supply, for example, secondary canal. In some cases it may be possible to raise the minimum water level, for example, by installing a cross regulator, or raising the top water level of a reservoir.

In practice the maximum practical height for the structure will depend on the materials and method of construction used, so long as the maximum hydraulic head is less than the pipe pressure rating. Surge or water hammer events which result in higher pressures would cause spilling from the structure.

To calculate the absolute maximum height for a pumpstand and header tank will require structural analysis of its detailed design, but an indication of the practical height limit is given in Table 4.2 for some of the types which have been widely built. The height given is above ground level and where elevated land near the water source is available a greater head rise could be achieved.

For pumped systems the limitation is often the increase in pumping costs and the height of the tower is kept to that required to provide the outlet operating pressure at the critical outlet, allowing for friction losses (typically 2 to 3m) in the pipeline.

The capital cost of the structure increases with height and for heights up to around 3m above ground level the increase is almost linear. Above this the cost of deeper and heavier foundations and walls to support the structure's self weight and the water column results in substantially higher construction costs.

Table 4.2 **Practical above ground heights of header structures**

Type of tank	*Practical height limit (m agl)*
Elevated tank with column support	5-6
Non-reinforced concrete pipe rings	3-4
Reinforced brick pressure tower	4-5
Steel riser pipe	8-10

(m agl) = metres above ground level

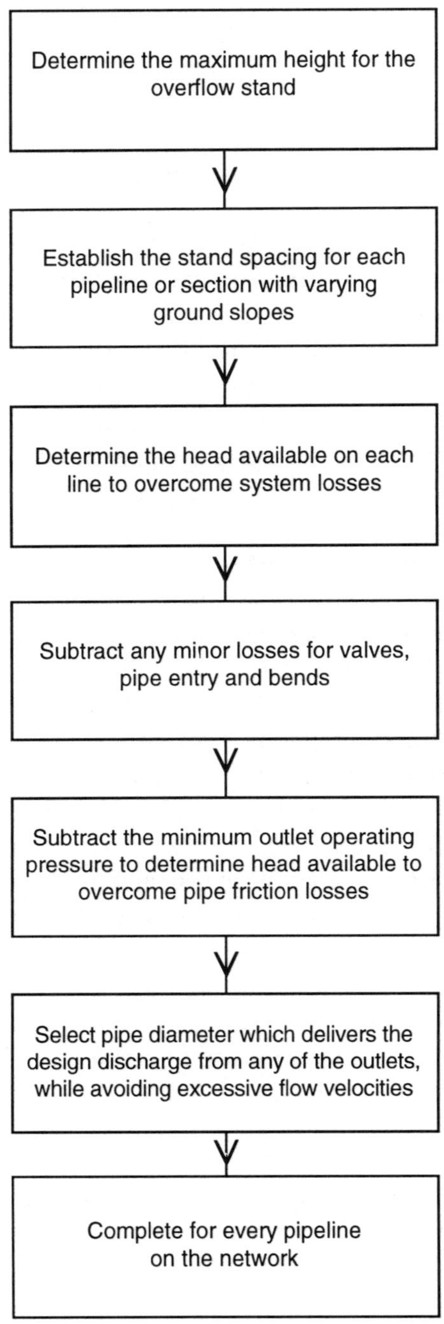

Flow chart 4.1. Hydraulic design of open pipe systems

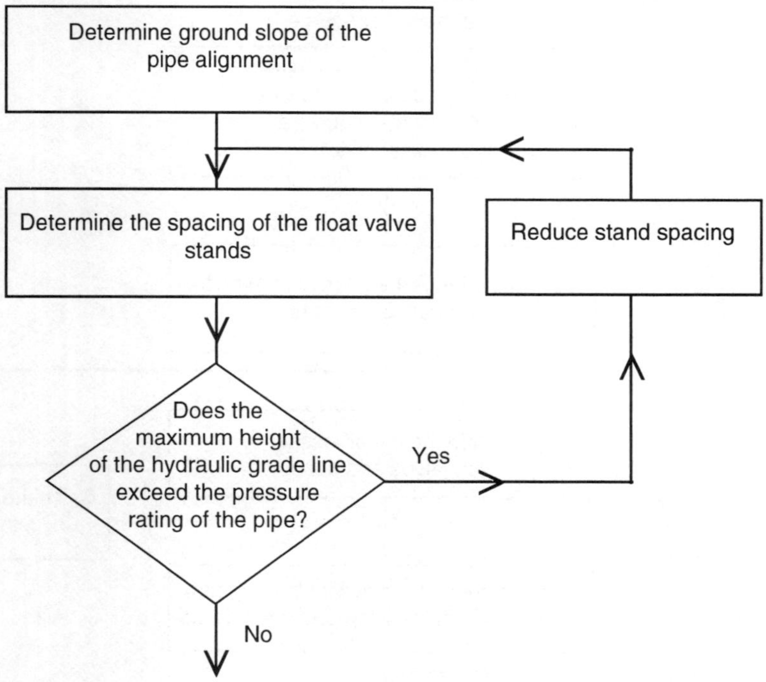

Steps

1. Select the pipe(s) with regard to maximum velocity of flow
2. Calculate the head to be dissipated at the float valve for the design flow rate
3. Select size and type of float valve, so that friction loss at design is less than head to be dissipated
4. Calculate the flow rate possible through the wide open valve
5. Determine the float setting for the design flow
6. Calculate the float submergence required to close
7. Calculate the maximum water level in the float stand
8. Determine stand height and maximum upstream pressure

Flow chart 4.2. Hydraulic design of semi-closed pipe systems (Design procedure for section of system upstream of a float valve)

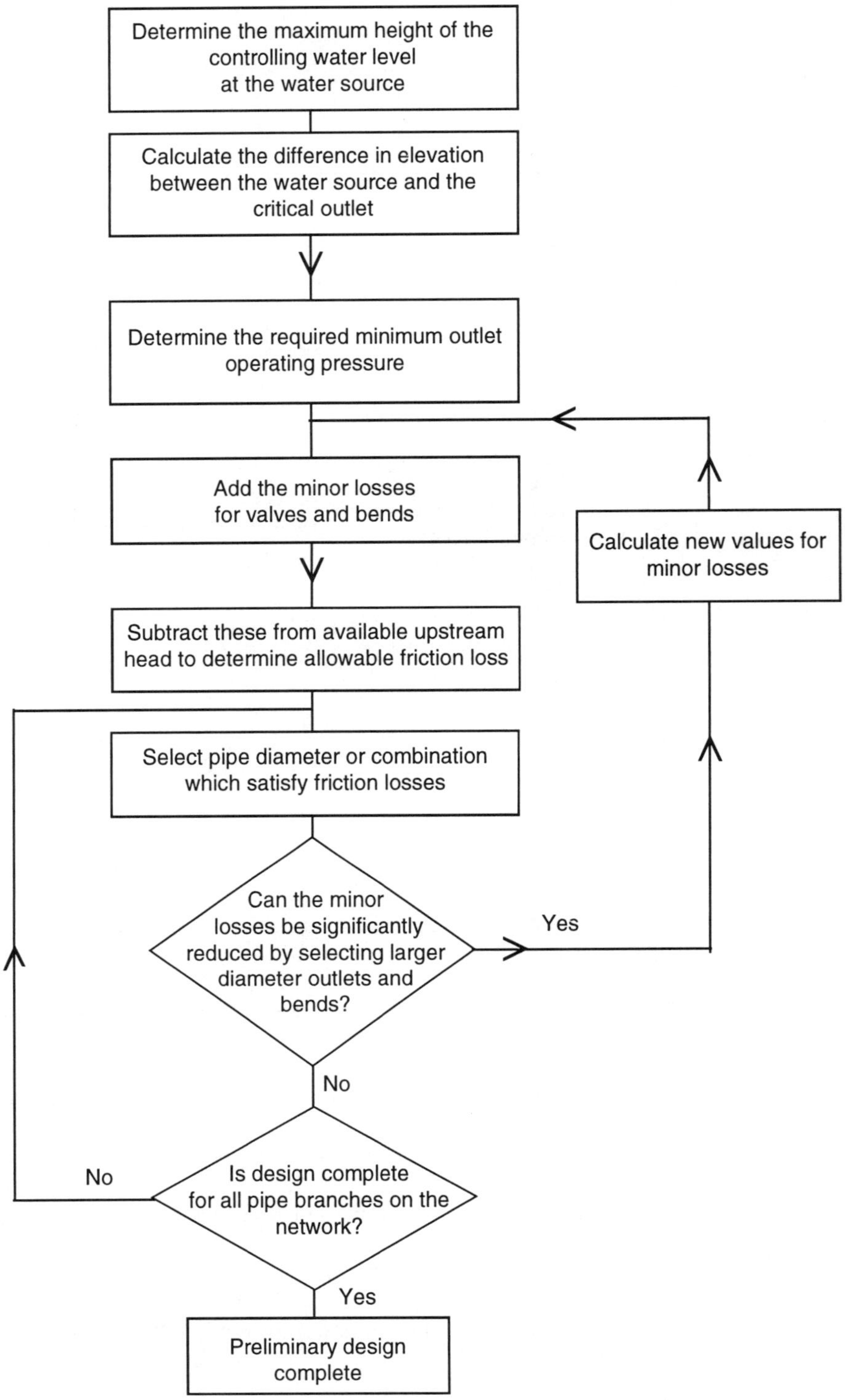

Flow chart 4.3. Hydraulic design of closed pipe systems

Outlet design water level

Sufficient head must be available at each outlet to allow irrigation of the furthest part of each outlet command area by whatever method of quaternary distribution is used. The design water level will in most cases be determined by outlets furthest from the tube-well and at intermediate outlets sufficient head should always be available.

As much as possible, outlets should be sited on or very near the highest point in the outlet command area to avoid having to raise earth channels on embankments and to ensure the widest possible choice of alternative methods of distribution.

The outlets and fields determining the design water levels for different parts of the scheme are often termed the "critical" outlets and fields.

The required outlet operating pressure or water level will depend on the following six factors:

Head loss through the outlet

Head losses through the outlet are normally restricted to less than 0.3m, and if high heads are to be dissipated the valve can be partially closed. High flow velocities require the provision of scour protection around the outlet and can result in excessive wear and vibration of the valve. Typically head losses through alfalfa valves will be in the range of 1.5 to 2.2 times the velocity head through the riser. Head losses through standard alfalfa outlets are shown in Table 4.3 (Booher, 1974). These head losses are based on precisely manufactured valves, and where local fabrication is involved allowance should be made for higher friction losses. Selection of the type of outlet and its size is outlined in more detail in Flow chart 5.1.

Head loss through the distribution structure

An outlet distribution structure will normally be associated with open field channels. Depending on the function of the distribution structure, friction losses will be associated with:

Table 4.3 Head losses through alfalfa valves (in millimetres depth of water)

Flow rate	*Diameter of opening (mm)*						
l/sec	*150*	*200*	*250*	*300*	*350*	*400*	*450*
10	20	10					
20	100	30	10				
30	260	70	30	10			
40	500	130	50	20	10		
50	830	220	80	40	20	10	
60	1260	340	120	60	30	20	10
80	2500	640	240	110	50	30	20
100		1050	400	180	90	50	30
125		1650	650	290	150	70	50
150		2500	970	440	220	120	70
175			1360	650	300	170	100
200			1850	830	410	220	140

Source: FAO (1974)

❍ the geometry of the structure;

❍ division of flow;

❍ flow across a weir and stilling basin.

Total head losses for an outlet distribution structure at design flow will normally be less than 0.2m and should not exceed 0.3m.

Head loss in the quaternary system to the critical field
The head loss will depend both on the quaternary system used and the length of the system to the critical field.

Although a range of quaternary methods can be used with low-pressure pipe systems including surface pipe, hose and open channels, the most common method is earth channels.

The distance from the outlet to the critical field in the case of earth channels is a balance between the maximum acceptable seepage losses and the higher investment costs of a pipe system with a greater outlet density. Although surface pipes and hoses can often take the most direct route from outlet to field, higher operating pressures are required to enable smaller diameter equipment of lower capital cost to be used. Hence the lengths of pipe and hose distribution systems will be limited by the available pressure and the capital cost of the equipment.

Selection of the maximum distance of the quaternary length from outlet to field has already been discussed in Chapter 3.

Typical operating pressure requirements of the outlet for the different methods would be as follows:

Earthen channel	0.2m
Gated pipe	0.2-2m
Surface pipe	0.5-2m
Layflat hose	1-1.5m

Head loss from the quaternary system to the field
The magnitude of this head loss will be small, depending on the way water is transferred from the quaternary system to the field. In pipe and hose systems it is included in the operating head for the system needed to overcome exit losses from one or more orifices. In open channels either of the following may be provided to direct water to the field:

❍ permanent or temporary outlets in the side wall of the channel;

❍ siphons.

Maximum depth of water ponded on field during irrigation
This will depend on the crop grown and soil type but is unlikely to exceed 0.15m.

Relative difference in ground level between outlet and critical field
The outlet should be higher in elevation than the critical field to avoid embanked earth channels. Where the outlet is lower than any of the fields, surface pipe or hose can be used to transfer outlet pressure as an alternative to building a raised earth channel. Conversely outlets which are too much higher than the surrounding fields require that drop structures or scour protection be provided.

The required outlet operating pressure or water level will be equal to:

Head loss through the outlet
+
Head loss through the outlet distribution structure
+
Head loss in quaternary system to critical field
+
Head loss from the quaternary system to the field
+
Depth of water ponded on the field during irrigation
+
Ground level of the critical field.

Minor losses

In general, unless poor selection of fittings and structure sizes has been made, or pipe assemblies are particularly short, system head losses due to bends and valves comprise only 5 to 10% of total pipe friction losses and are frequently referred to as minor losses. Unnecessary bends and fittings should however be avoided as not only do they add to the friction losses but they are also more likely to be positions at which leakage occurs.

Head losses are usually expressed in terms of velocity head:

$$H_m = \frac{K_m V^2}{2g}$$

Equation 4.1

where K_m = friction loss coefficient (also known as a form loss coefficient)
V = velocity before the feature causing friction loss (m/s)
g = acceleration due to gravity (m/s^2)
H_m = minor head loss (m)

For the situations encountered on buried pipe systems the following friction loss coefficients are considered typical:

90° bend	0.2 to 0.3
45° bend	0.1 to 0.15
22.5° bend	0.05 to 0.1
90° tee in-line	0.1
90° tee side outlet	1.3 to 1.8
Pipe into stand	1
Pipe out of gate stand	0.5 to 0.8
Alfalfa valve outlet	1.5 to 2.2
Alfalfa inlet	4 to 6

Values for a wider range of types of valves and situations which may be encountered on buried pipe distribution systems are detailed in Table 4.4.

Table 4.4 **Friction loss coefficients for buried pipe systems**

Feature	K_m
1. Tank to pipe	
pipe projecting into tank	0.8 to 1
pipe flush with wall	0.4 to 0.5
small radius at orifice	0.1 (radius < 4D)
large radius at orifice	0.05 (radius > 4D)
2. Pipe to tank, all cases	1
3. Bends: r = bend radius, D = pipe diameter (US National Bureau Standards)	
90° r/D = 1	0.35
2	0.18
5	0.20
10	0.32
15	0.38
20	0.42

For 60° bend: K_m = 0.83 x value above for particular r/D,
45° bend: K_m = 0.7 x value above,
and 22.5° bend: K_m = 0.42 x value above (USBR)

4. Valves (full opening)	
angle	2.5 to 5
globe	10
swing check (reflux)	0.6 to 2.5
gate : 300mm pipe	0.07
: 50mm pipe	0.16
needle	0.5
butterfly	0.3
float balanced (double disc)	1.9
float unbalanced (single disc)	2.4
5. Tee,	
through side outlet	1.8
through side inlet	1.3
straight through	0.1
elbow, short	0.9
medium	0.8
long	0.6
45°	0.4
return	2.2
square	1.27

Note: manufacturers' catalogues should be consulted

6. Expansion $h_m = K_m V_1^2/2g$
Sudden $K_m = (1-D_1^2/D_2^2)$
Gradual K_m from table below, = total cone angle

θ°	D_1^2/D_2^2 0	0.2	0.4	0.6	0.8	1.0
7.5°	.13	.08	.05	.02	0	0
15°	.32	.24	.15	.08	.02	0
30°	.78	.45	.27	.13	.03	0

Contraction, sudden $h_m = K_m V_2^2/2g$

D_1^2/D_2^2	0	0.2	0.4	0.6	0.8	1
K_m	0.5	0.37	0.25	0.15	0.07	0

where V_1 and D_1 are initial values

Source: Labye et al (1988)

Pipeline design

Optimization of pipe size

In complex pipeline systems, because pipe cost increases with pipe diameter, and pumping costs reduce with pipe diameter, pipe selection is often carried out using an optimization process. On the simple pipeline networks considered here, the range of pipe sizes and materials will be limited, so that selection by trial and error will not usually be onerous. Standardization of pipe sizes and materials can reduce design and construction costs, so outweighing the benefits of precise selection from a wide range of pipe options. Where local manufacture of pipe is being considered, selecting a limited number of pipe sizes clearly reduces the investment required; for example, moulds for spinning of non-reinforced concrete pipe.

If a large number of schemes are to be built and they have similar sizes and features in their command areas, careful selection of a small number of sizes of pipe may suffice.

Where a wide range of pipe sizes is available — for example, from a central supplier — optimal matching of the head available with the pipe friction losses is made simpler.

Because the pipe material costs represent from 60 to 90% of the cost of the pipe system depending on which pipe material is used, designers try to select the minimum capital cost combination of pipes which delivers the required flow.

Before selecting the pipe sizes, two parameters need to be set for the whole or different parts of the pipe network:

- the available head either from a gravity or pumped source;

  Higher flow velocities may help reduce pipe size but lead to increased head requirements at the inlet. Lower flow velocities as well as reducing pumping costs also mean less variation in discharge from outlets operating simultaneously on the same closed pipe network.

- the flow capacity of the pipeline;

  Larger diameter pipes can provide increased operating flexibility which may outweigh their higher capital costs.

Where the supply flow rate is determined by a pump, the flow capacity is sized according to the pump discharge. Where a pipe system is fitted to an existing water source and pump installation, the flow capacity should reflect any expected improvements to the yield of the water source, or any possible changes to the operating requirements of the system. For example, a pump is likely to be replaced several times in the life of the distribution system. Changes to the cropping pattern and irrigation methods practised can mean requirements for higher flow rates.

In practice, for the simple pipe network considered, minimizing the capital cost of any pipe section will usually involve not more than two different-sized pipes and probably 90% of the different sections will comprise only a single pipe diameter.

Loop or branching systems

On loop pipe systems the aim should be to select one pipe size for the entire loop. Where a large range of pipe sizes is not available, two sizes of pipe may be required to provide the required balance between head loss and capital cost.

On branching systems pipe size will tend to decrease away from the water source. The critical branch and outlet will determine the selection of pipes from the range available, given the head available to overcome friction losses. Strict optimization of pipe selection would allow smaller diameter pipe to be installed on short non-critical pipe branches.

However, in general pipe lengths of less than 50m are not fitted with a second smaller diameter pipe, as small quantities of pipe can be relatively expensive to procure and construct.

Using larger pipe than required for some of the network may allow a slight reduction in the operating cost where a pumped source is involved. Whether using larger diameter pipe than required on the non-critical pipe branches will reduce the operating costs where a pump is involved depends on the selection of the operating and control system. With an elevated tank or top water level control pressure tower, the delivery head will largely be fixed by the weir or tank height.

Pipe friction estimates

The most important information relating to the design of the pipe system is awareness of the peak flow rates which must be conveyed in different parts of the system. A system which has insufficient capacity will be severely limiting and expensive to rectify. Increasing flow velocity will not usually be a viable alternative for low pressure pipes. In general for the pipe materials and pressures involved flow velocities should seldom exceed 1.5m/s.

Of all the equations used for the estimation of pipe friction losses the Colebrook-White equation, combining the theory of turbulent flow with experimental data for flow in commercial pipes, appears the most reliable. It covers the whole range of turbulent flows whilst allowing for temperature. Equations suitable for hand-held calculators have been developed which enable a pipe to be designed given any three of the four variables, roughness height (k), pipe diameter (d), flow velocity (V) and friction slope (s).

Tables and nomographs have been developed by Hydraulics Research (1983 and 1990) for a wide range of the variables and covering all the situations likely to occur in buried pipe distribution systems, simplifying calculation. Field measurements confirm the general accuracy of the equation when representative values of roughness height are chosen.

The choice of roughness value should take into account not only the quality of the jointing but also the variation in surface roughness found in pipes which are considered to be the same diameter.

Values based on field-verified measurements are shown in Table 4.5 below. These are conservative and manufacturers may quote more optimistic values, for example k = 0.015 for uPVC pipe. For the small pipe sizes and low flow velocities generally encountered this will make only a small difference to the estimated friction loss. In situations where unknown pipe materials or pipe system designs are to be used, field measurement must be carried out early to verify the choice of roughness height.

Nomographs derived by Hydraulics Research are included for normal values of roughness height to enable friction loss estimates to be made (see Figures 4.2, 4.3 and 4.4).

Table 4.5 **Recommended values of roughness height (k)**

Condition of pipe	*uPVC*	*Non-reinforced concrete*	*Asbestos cement*
good	0.03	0.3	0.015
normal	0.06	0.6	0.03

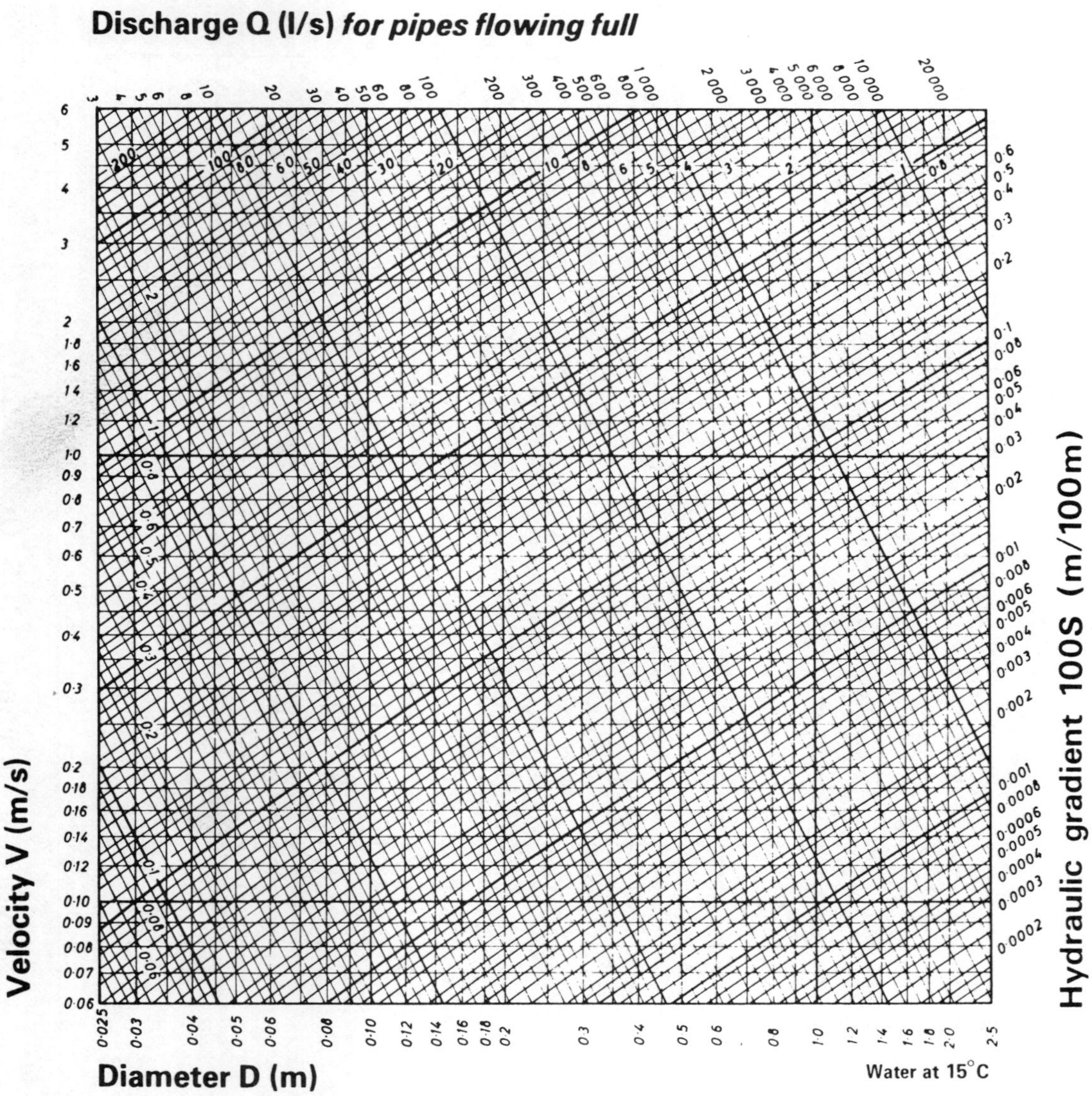

(Non-reinforced concrete pipe in normal conditions with k_s=0.6mm)

Figure 4.2. Chart for hydraulic design of pipes

Source: Hydraulic Research (1990)

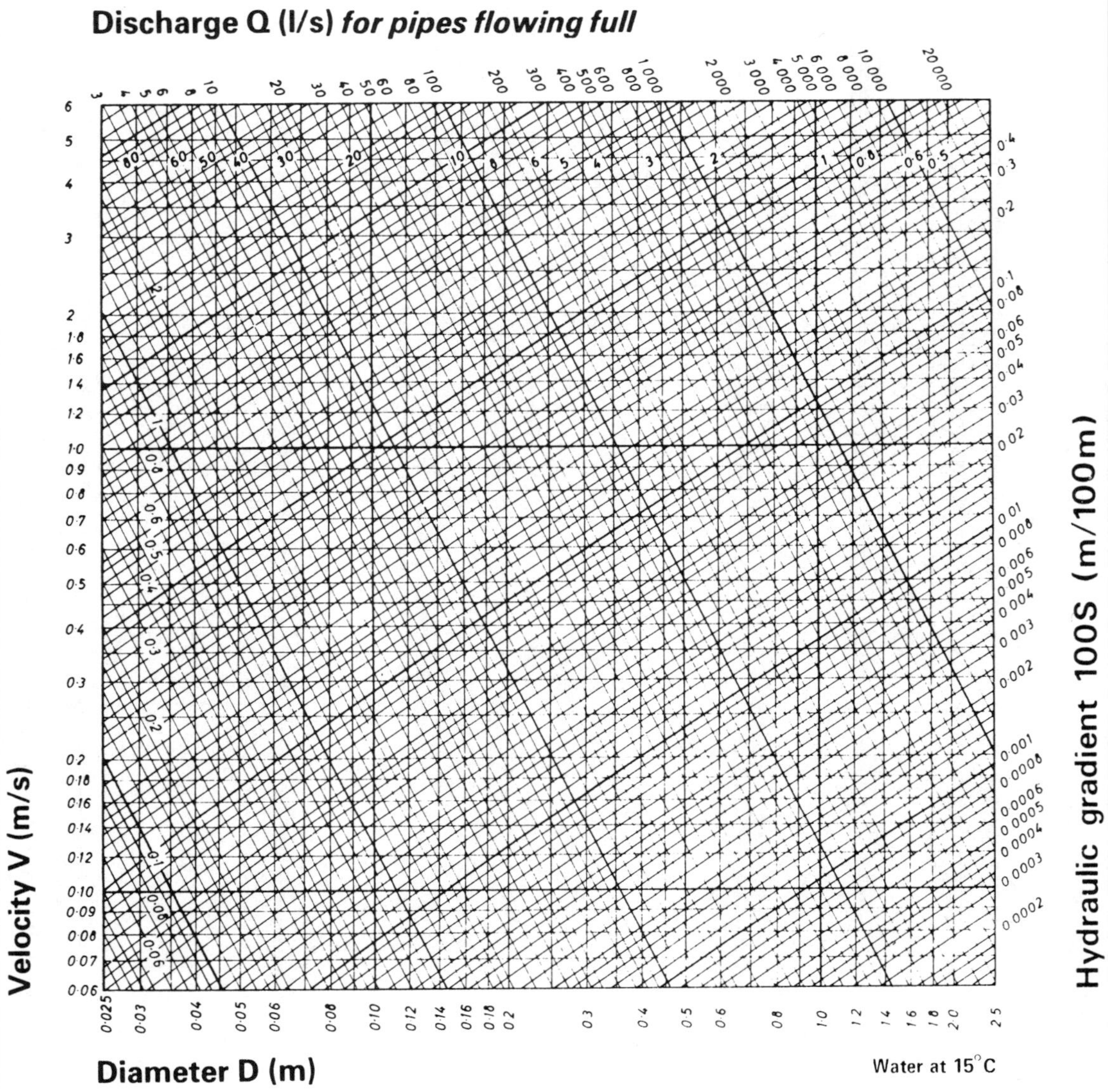

(asbestos cement pipe in normal conditions with k_s=0.03mm)

Figure 4.3. Chart for hydraulic design of pipes

Source: Hydraulics Research (1990)

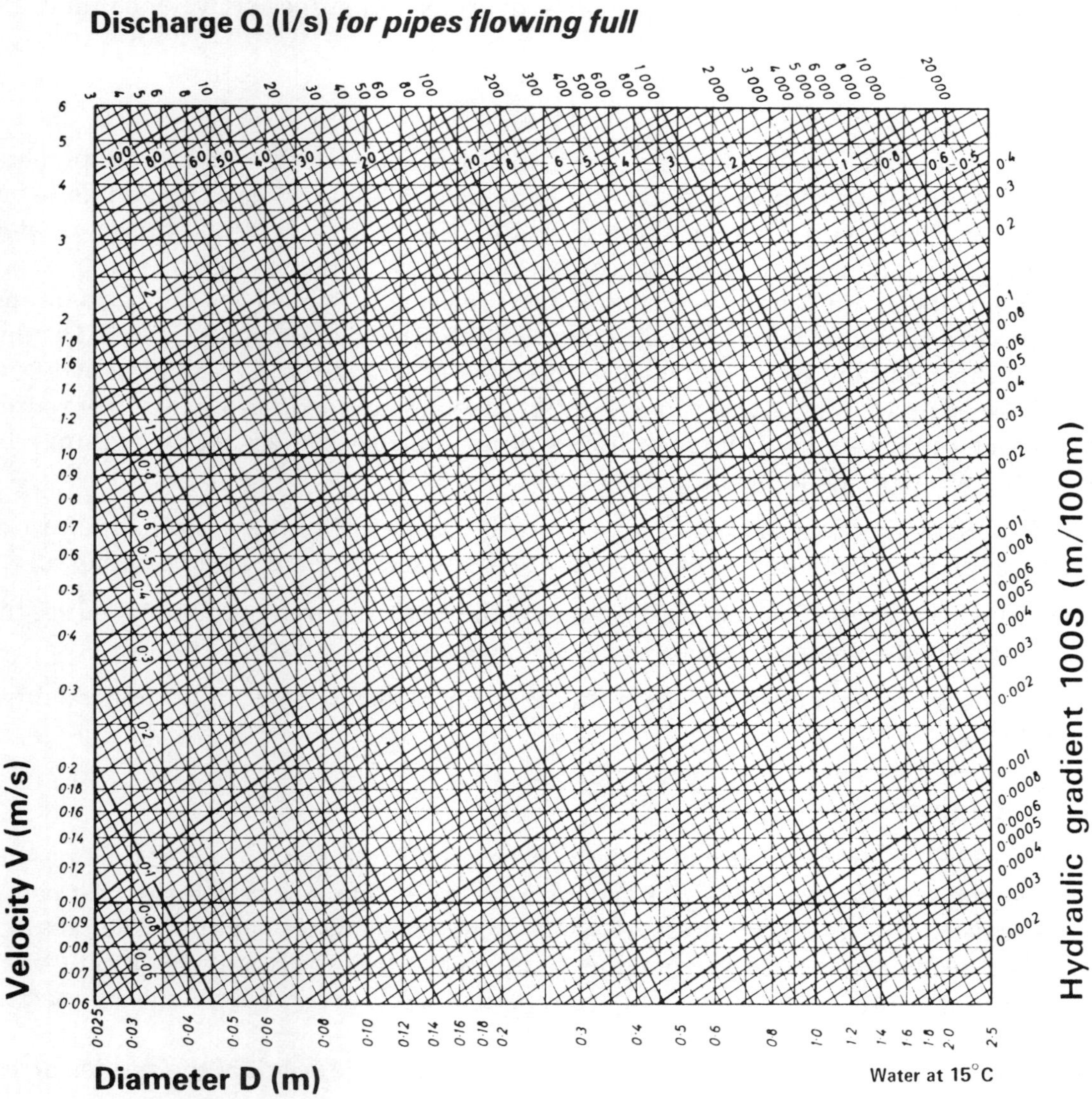

(uPVC pipe under normal conditions with k_s=0.06mm)

Figure 4.4. Chart for hydraulic design of pipes

Source: Hydraulics Research (1990)

Surge and water hammer protection

In this section the requirement for and design of protection for water hammer and surge events is outlined. Particular attention is given to closed pipe systems where open stand-pipes can be installed to minimize the magnitude of the pressure rise.

Before discussing the design and provision of protection, surge and water hammer are briefly defined.

Surge

Surge is considered to be any transient pressure fluctuation which may occur in pipeline systems at atmospheric pressure. During surge, water flow is characterized as being unsteady, oscillating from one steady state condition to another. Surge is often associated with the entry and exit of air which becomes trapped in the pipe system.

Surge problems in low-pressure pipelines are usually caused by the sudden and uncontrolled release of entrapped air from the line. The air may have collected at one position during filling or be entrained in the flow at structures such as pipe overflow stands. Open pipe systems, where overflow stand-pipes are used, suffer more from surge than other pipe systems. If large volumes of air are released a shock wave may be generated. Ways of avoiding surge include:

- provision of air valves or vents in pipe systems to provide for the controlled release of air (the selection and design of air vents are discussed in detail late in Chapter 5);
- avoiding severe and unnecessary changes in the vertical pipe alignment to reduce the positions in the pipeline system where air is trapped for later sudden release;
- ensuring that pipe systems remain full of water, which dramatically reduces surge problems.

Water hammer

When the kinetic energy of moving water is transformed into pressure energy, a pressure wave is generated that oscillates back and forth in the pipeline. At any point in the pipeline this is registered as a surge in pressure which is known as water hammer. The pressure wave is reflected back on itself when it encounters any free water surface, usually at an open stand structure, and becomes superimposed on itself so resulting in the dampening of the surge.

Water hammer in buried pipe systems usually occurs as a result of the sudden arrest of flow caused by one of the following:

- sudden valve closure;
- sudden release of air;
- sudden stoppage of a pump.

The sudden release of air can be avoided by adequate provision for air release using valves or vents. In the case of sudden pump stoppage and valve closure the surge in pressure can be damaging if not restricted to below certain limits. The hazards include:

- If the total of the initial pressure and the excess pressure developed at water hammer exceeds the maximum permitted pipe pressure, then there is a risk of the rupture of the pipe or joint.
- If negative pressure results then a pocket of cavitation may be created, causing possible collapse of the pipe wall (particularly in plastic pipes) or ingress of pipe seals.

- Cycling positive and negative pressures may lead to acceleration of pipe failure over time, due to pipe fatigue.

The requirement for surge and water hammer protection and the choice of methods of protection available depend on the type of buried pipe system involved.

Open and semi-closed pipe systems

In open and semi-closed systems the spacing of overflow stands and float valve stands usually ensures that water hammer events are quickly dampened in the stand structures, restricting the size of the pressure rise. For rapid valve closure, in most cases the spacing of the stands is such that outlet valves are well within the minimum distance from a free water surface. The method for calculating this minimum distance is defined below and in Flow chart 4.4.

Where pipelines are flat and control stands infrequent, reference should be made to the design for protection on closed systems, as additional open vents or pressure relief valves may be required. In systems where a number of irrigators are taking water at the same time, then changes in flow momentum are often small and provide few problems.

Problems with surge can usually be avoided by sensible methods of operation particularly during pipeline filling. Methods which will assist in avoiding the entrapment of excessive quantities of air and its subsequent sudden release include:

- provision of undershot gates in overflow stands in open pipe systems (Merriam, 1987a);
- slow and careful filling of the pipe system from empty.

Closed pipe systems

In closed pipe systems an analysis of the need for water hammer protection should always be made. The choice of strategy to limit the extent of the pressure rise will depend on whether open vent surge risers can be installed. On most low-pressure pipe systems, this is usually possible. Where this is not possible special pressure release valves would need to be used. Use of pipe with a higher pressure rating can avoid the need for these structures.

Sudden pump stoppage

In most pipe systems with a pumped supply an open stand-pipe or tank is provided between the pump and the pipeline, and is adequate for protection of the pipe network from water hammer due to sudden stoppage of the pump. Where the pump is directly coupled to the pipeline then a pressure release valve will be required.

Sudden release of air

Water hammer due to sudden release of air in the pipeline is prevented by the installation of air vents or release valves at high points or problem areas. Any open stand structure will function as an air vent and reduce the likelihood of sudden air release occurring.

Sudden valve closure

This constitutes the predominant cause of water hammer in closed pipe systems, and minimizing the pressure rise which occurs is the object of design for open vent surge risers.

Methods for reducing the damaging pressure rise due to sudden valve closure

The following section is largely based on Campbell (1986). The analysis of water hammer is complex and any attempt to predict the conditions arising is necessarily an approximation. However, in general, if the time of valve closure (T_1) is more than 20 times the instantaneous closure time (T_c) then surge pressure can be regarded as insignificant. The instantaneous closure time is considered to be the time taken for the pressure wave developed at closure to travel the round trip distance to the nearest free water surface.

Instantaneous valve closure is calculated using Equation 4.4 on page 98. For any situation in between, that is $T_c < T_1 < 20\,T_c$, then design decisions need to be taken to reduce either the pressure rises or their damaging effects .

The magnitude of the pressure rise can be controlled in a number of ways, including the following:

Control of the rate of valve closure by:

- use of valves with finely threaded spindles so that more turns are required for the same degree of valve closure: most of the alfalfa valves in common use are manufactured with this in mind;
- use of valves opened with a special spanner which must be retracted and repositioned for each proportion of a turn. Campbell (1986) mentions an outlet design of this type which was to be used on the Kallada Irrigation Scheme (see Figure 4.5).

There is little reference to the valve closure periods achieved in practice with these methods. Field experience suggests that valve closure periods in excess of 20 seconds may be difficult to achieve.

Installation of surge risers

The selection and spacing of surge risers is the primary method for limiting the magnitude of the pressure rise, and the steps involved in design are outlined below. Surge risers will be necessary on all pipe systems although in practice any open stand structure of adequate diameter will serve to reflect the pressure wave.

It will be important to analyse the magnitude of the rise in pressure with more rigid pipe materials such as concrete and asbestos cement.

Installation of pressure relief and vacuum relief valves

Though valves of this type have an application in closed pipe systems they are normally used in pump-supplied systems where pump stoppage and power failure are common causes of water hammer. Simple pressure relief valves which can be actuated at relatively low pressures are not available.

Care during pipe installation

It is important to keep the hydraulic grade line above the elevation of the pipe. If the pipe approaches the hydraulic grade line then clearly pressures below atmospheric become a possibility under the negative pressure waves produced after valve closure. Keeping the pipe grade low may necessitate avoiding intermediate high spots along a pipe section or providing for a deeper excavation cut. Vacuum valves could be installed but the air they admit during negative pressures could later cause surge problems if suddenly released.

Design procedure for closed pipe systems with open vents

The usual approach must be to specify pipe which can withstand the estimated pressure surges under as slow a valve-closing regime as is practicable and provide surge risers where required. The design procedure set out is drawn from Campbell (1986).

For the purposes of pipe system design an estimate must be made of the pressure rise associated with the expected minimum period of valve closure or the occurrence of pump stoppage. This pressure surge when added to the maximum operating pressure at that position must be less than the pressure rating of the pipe. The analysis is repeated for a number of critical positions in the pipe system.

Flow chart 4.4 outlines the procedure for estimating the maximum surge pressure and determining the minimum distance from a valve to an open stand structure, which is detailed below.

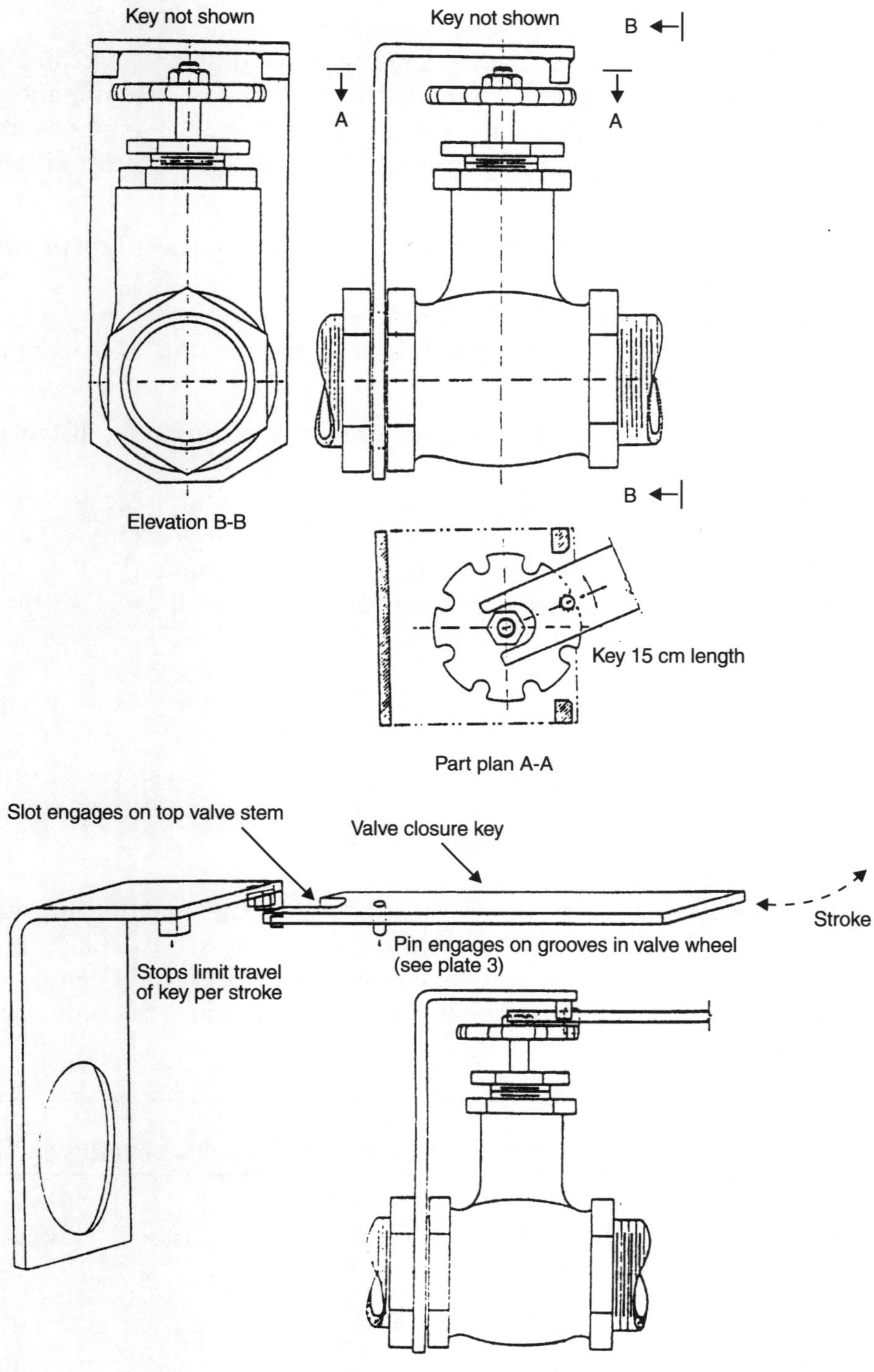

Figure 4.5. Attachment to gate valve to ensure slow closure

Source: Campbell (1986)

Estimating maximum surge pressure after sudden valve closure

Step 1

Calculate the maximum allowable surge pressure rise.

This will be defined by the maximum operating pressure at that point and the pressure rating for the pipe. As previously mentioned for uPVC pipe, surge pressure should not exceed 30% of the pipe pressure rating. For low-pressure concrete pipe, surge pressure added to operating pressure should not exceed 25 to 35% of the hydrostatic test pressure.

Step 2

Select a first-guess estimate for the maximum distance between the site of pressure rise and any free water surface.

For uPVC pipe systems this would be from 600 to 800m.

For non-reinforced concrete pipe systems this would be from 300 to 500m.

Step 3

For the maximum flow calculate the peak pressure rise occurring at the instantaneous time of closure.

Peak pressure rise occurs when valve closure or pump stoppage occurs in any time period which is equal to or less than the time it takes for the pressure wave to travel the round trip to a free water surface (called the time of concentration).

Peak pressure rise and negative pressure drop in metres will be given by

$$\frac{V_o \times U}{g} \qquad \textit{Equation 4.2}$$

where V_o = velocity of flow before valve closure or pump stoppage (m/s)
U = pressure wave velocity in the particular pipe material (m/s)
g = acceleration due to gravity ($9.81 m/s^2$)

The wave velocity is a function of the elasticity of the water and the pipe wall material. The wave velocity, however, does not vary significantly over a range of pipe sizes provided the material is the same. The wave velocity cannot exceed 1400m/s which is the speed of sound in water. Typical values for wave velocities in commonly used materials are listed in Table 4.6.

Table 4.6 **Surge pressure wave velocities in different materials**

Pipe material	*Wall thickness (mm)*	*Pipe diameter (mm)*	*Ratio*	*Wave velocity (m/s)*
non-reinforced concrete	25	250	0.1	1100
non-reinforced concrete	25	300	0.083	1060
uPVC ($2.5 kg/cm^2$)	4.5	250	0.017	220
uPVC (thin-walled)	2.5	25	0.01	170
steel	4.5	250	0.017	1140

Note: The reciprocal of the ratio of wall thickness to pipe outside diameter is known as the dimension ratio (DR).

Source: MMP (1989a)

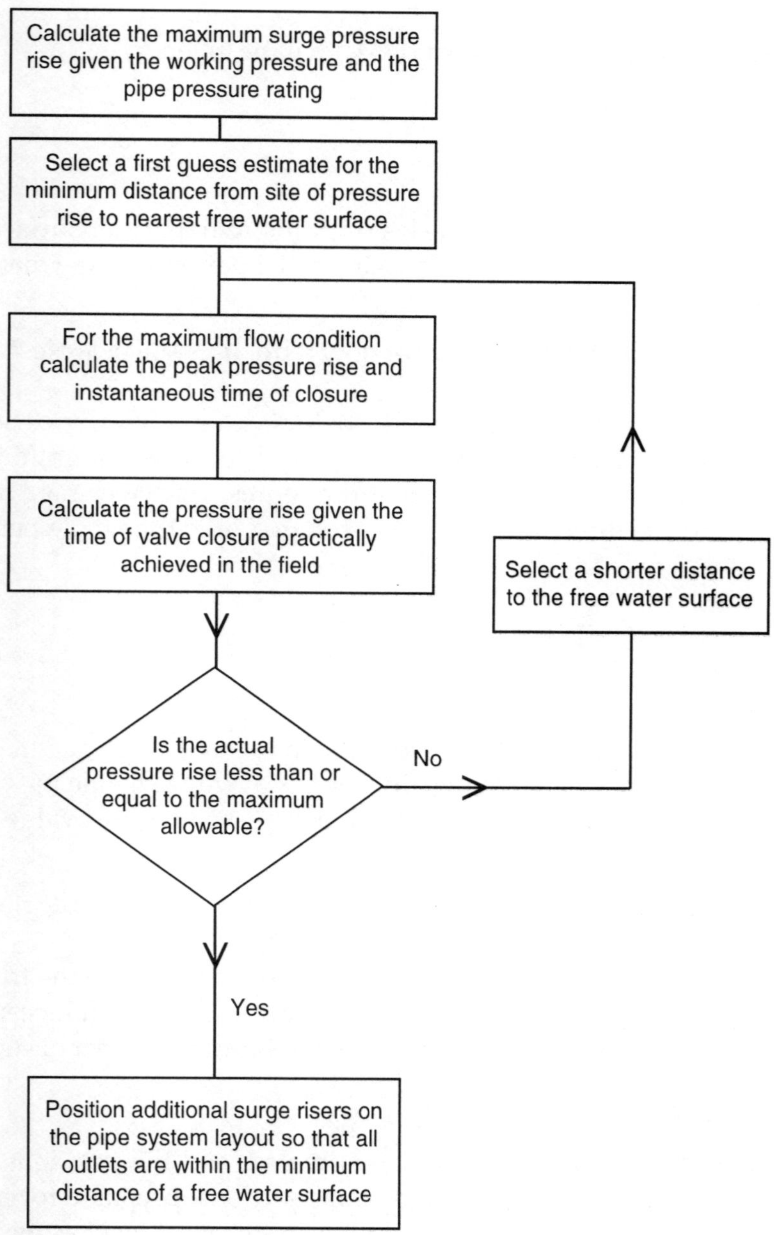

Flow chart 4.4. Water hammer protection for closed pipe systems with open vents

The value for the negative pressure drop or rise in pascals is represented by

$$P\ (KPa) = W \times g \times P\ (m) \qquad \textit{Equation 4.3}$$

where W = density of water (1 tonne/m^3)
g = acceleration due to gravity (9.81m/sec^2)

The time of concentration in seconds (T_c) will be given by

$$T_c = \frac{2L}{U} \qquad \textit{Equation 4.4}$$

where L = distance from valve or pump to free water surface or reservoir (m)
U = pressure wave velocity in the particular pipe material (m/s)

Step 4
Calculate the pressure rise for rates of flow reduction which are less than at instantaneous closure.

In order to estimate the pressure rise for longer times of valve closure an approximation must be made because the rate of change of velocity is not uniform over the period of closure. The assumption made is that the representative change in velocity during the round trip time is three times the rate averaged over the whole period of closure.

Thus average change in V_o in T_c

$$= \frac{0.5\ T_c \times V_o}{T_1} \qquad \textit{Equation 4.5}$$

where T_1 = the time of valve closure in seconds
Representative change in $V_o = 3 \times$ average change in V_o.
Maximum head rise is given by substituting this value in Equation 4.2.

Step 5
Compare the calculated value of pressure rise with the acceptable maximum surge pressure.

If the surge pressure is too high then a shorter distance to the free water surface must be selected. If the surge pressure is substantially lower than the acceptable value, a greater distance should be selected so that only the minimum number of surge risers is provided.

Pipeline long section

When the frictional losses have been estimated, and the positions of structures such as surge risers and air vents determined, a long section is prepared for each section of the pipe network showing the hydraulic gradeline and ground level as well as the profile of the pipeline at its recommended embedment depth. An example of such a long section is shown in Figure 4.1.

Although it may be possible to prepare the long section using survey data already collected, a more detailed survey may be required along the selected pipeline route.

From the long section a check should be made on the maximum height of the hydraulic grade line, and that it is within the limits of both maximum operating pressure and the maximum physical height of the structures. Guideline maximum operating pressures (including an allowance for surge) are given in Table 1.4 and will depend on the pipe material used. The hydraulic grade line is used to set the height of structures (within their physical limit).

CHAPTER 5

DESIGN CONSIDERATIONS FOR COMPONENT STRUCTURES

Introduction

Inlets, pumpstands or header tanks

Inlets, pumpstands and header tanks comprise the full range of structures which connect the pump or gravity supply with the pipe system. The structures are divided into several different types depending on the nature of the pump connection and the way in which the driving head for the pipe system is provided. For hydraulic design of the buried pipe system, two broad groups are distinguished:

- Structures where water is forced to rise in a tower or pipe to a level above the inlet from the pump to that required to overcome the system friction losses. Pressure towers, header tanks and riser pipes fall into this group of pumpstands.

- Structures where the pump or gravity water supply discharges freely into a stand or tank. The height of the discharge pipe or outlet sets the available driving head. Low head concrete pumpstands and elevated tanks fall into this group as well as inlets from gravity water sources. A typical low-head pumpstand is shown in Figure 5.1 (US SCS, 1967).

Design of individual pumpstand and header tank structures is considered below for both groups, followed by discussion of the design of gravity inlet structures.

Pumpstands and header tanks

The pumpstand or header tank is designed to provide a free water surface at the level of the hydraulic grade line at the head of the pipe system. Freeboard of a minimum of 0.3m is added to give the design height of the structure.

The friction loss analysis has already taken into account the practical height of the pumpstand or header tank in estimating the available head. If the final design head is outside the limit of the gravity supply or the type of header tank envisaged then the design must be repeated with a lower value of available head.

In designing header structures the following should be remembered:

- The water source should be as close as practicable to the higher land which is to be irrigated.

- For a new tube-well installation, it will be generally cheaper to construct a deeper well on higher ground than to build a taller header tank. If the well is sited on a high spot then the greater head available may make smaller diameter pipe an option, so lowering the pipe system capital costs.

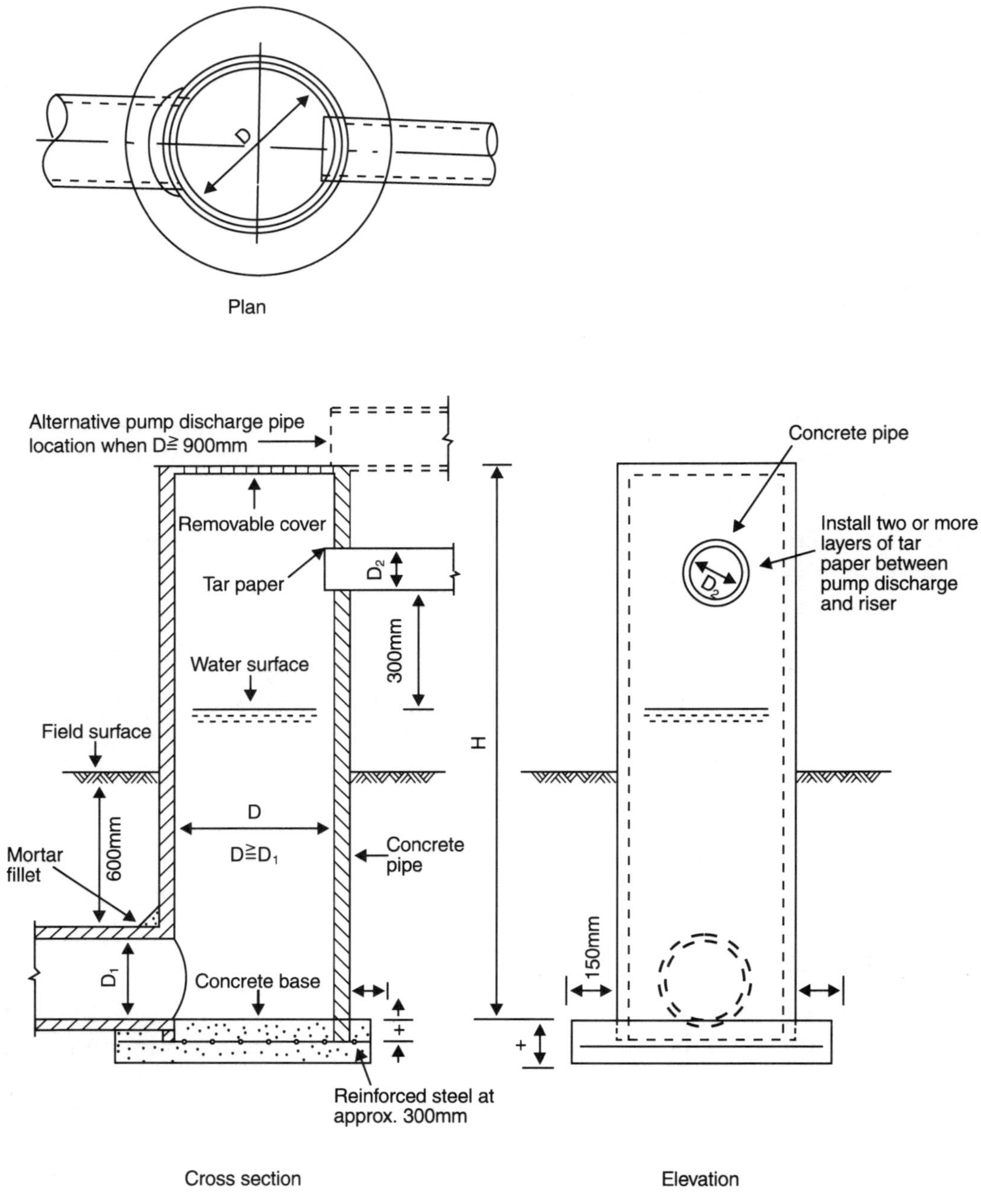

Figure 5.1. Low head pumpstand

Source: US SCS (1967)

American Society of Agricultural Engineers, recommendations on the design of pumpstands (Koluvek, 1970) state that:

- The centreline of the pump discharge pipe should have a minimum offset from the centreline of the outlet pipe equal to the sum of the diameters of the inlet and outlet pipes.
- Check valves should be used in the pump discharge line wherever the potential backflow from the pipeline would be sufficient to drain the pipeline or damage the pump.
- Vibration from the pump discharge pipe must be isolated from the stand. Techniques used to achieve this include:
  - i) installation of a flexible joint in the pump discharge pipe, between pump and stand;
  - ii) use of flexible packing such as rubber inner tube, or bitumen-covered paper in pipe/stand joint.
- If the size of the stand is decreased above the pump discharge pipe, the top vent portion should have a cross-sectional area such that if the entire pump flow were discharging through it, the average velocity would not exceed 3m/s.

Head riser pipes should be equivalent to or greater in diameter than the underground pipe to which they are attached.

Pressure tower or header tanks should be of a size adequate to allow access to any valves and gates installed within the structure. Typically a minimum inside dimension of 0.9m will provide for access for maintenance. In a header tank with internal satellite towers a larger minimum size should be provided if any repair to the satellite stands is envisaged.

A pressure tower with bottom valve control as built in Bangladesh is shown in Figure 5.2. This is probably the most common design, and has been used in a number of situations. Figure 5.3 is of a header tank with top water level control via external satellite stands. This design has been used in at least one location in Bangladesh by IDA DTW II project (MMI, 1992), and similar header tanks have also been built by TADP. Examples of each of these types of structure are illustrated in the publication 'photographs of Buried Pipelines for Surface Irrigation'. The tank in Figure 5.4 uses an internal pipe arrangement.

The storage volume in a typical pressure tower is usually insufficient to provide buffer storage for automatic pump regulation using floats, without excessive pump cycling. Much larger structures must be provided for this, such as the elevated tank used on pipe systems built in Uttar Pradesh. Examples of this tank are illustrated in the publication 'Photographs of Buried Pipelines for Surface Irrigation'.

Gravity inlet structures

Some form of inlet structure is required where water is supplied from an open channel or reservoir.

The simplest of these is a gated orifice connecting the canal with the pipe system. A slide or screw gate provides a means of shutting off and regulating the inflow discharge to the pipe system. In open pipe systems the gate setting determines the inflow discharge, while for closed and semi-closed systems, inflow discharge will be set by the operation of the outlets on the pipeline.

Where the head at the inlet to the pipe system varies widely, some form of automatic inflow control structure may be required.

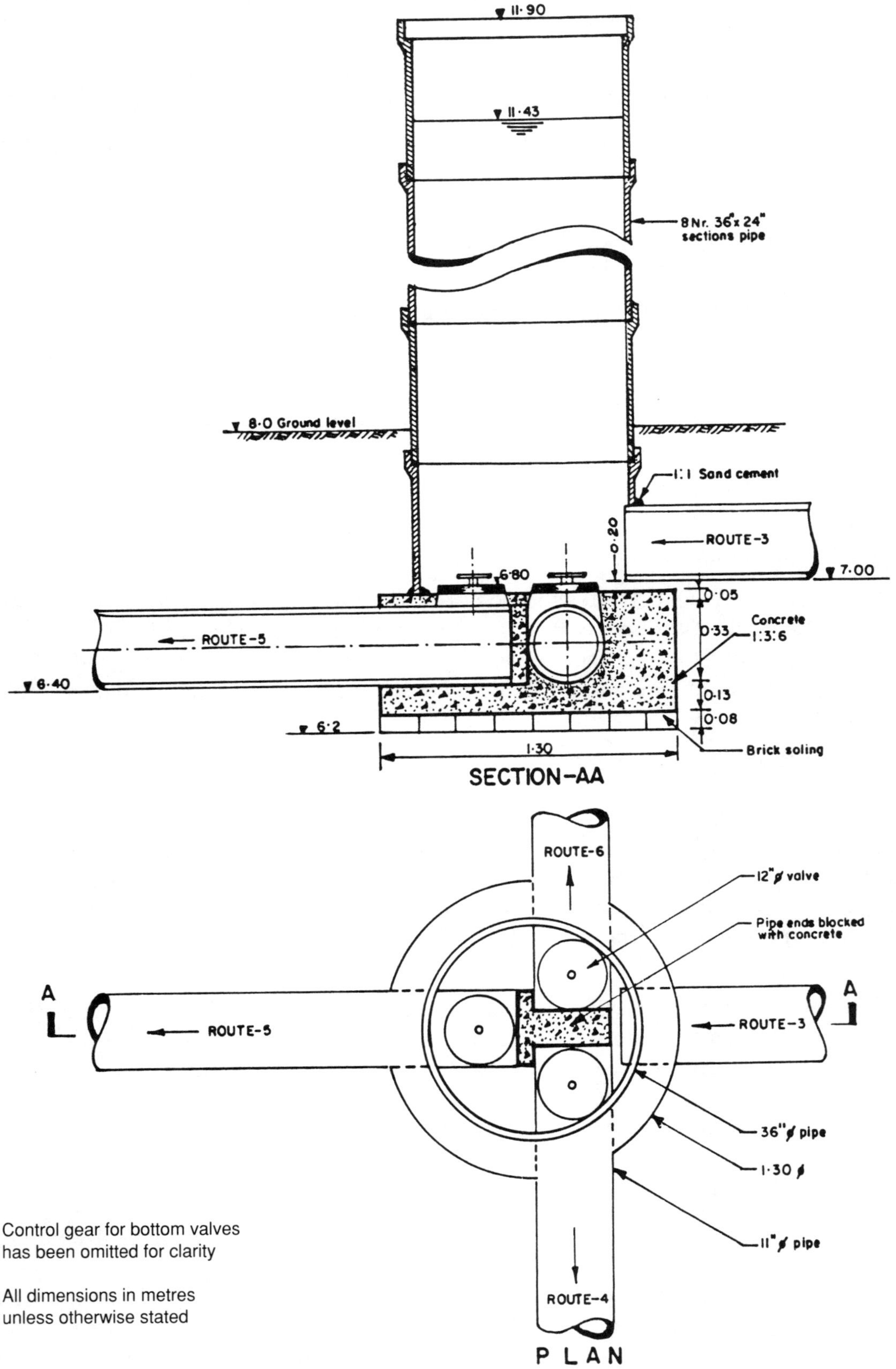

Figure 5.2. Buried pipe division tower

Source: MMI (1992)

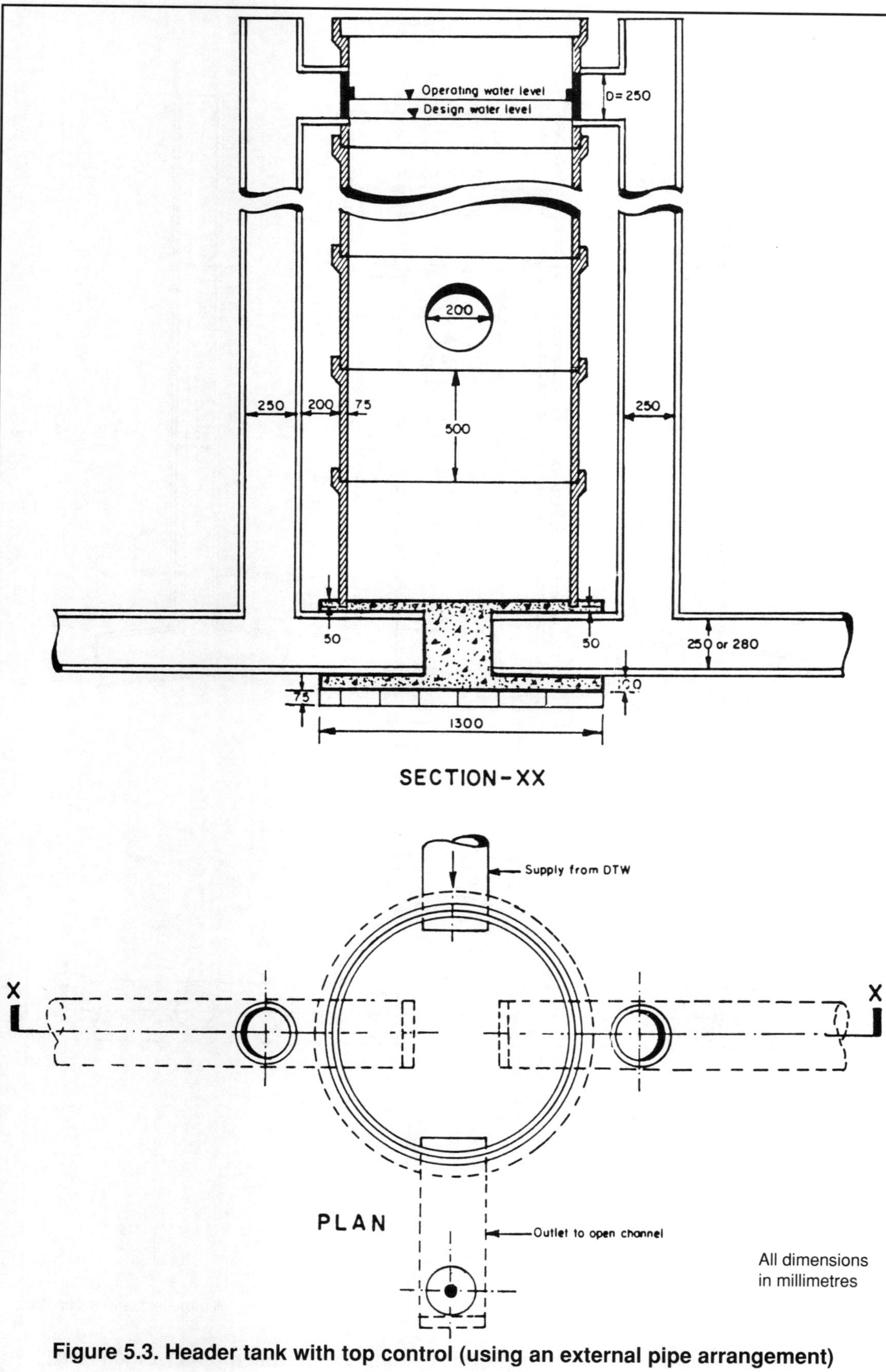

Figure 5.3. Header tank with top control (using an external pipe arrangement)

Source: MMI (1992)

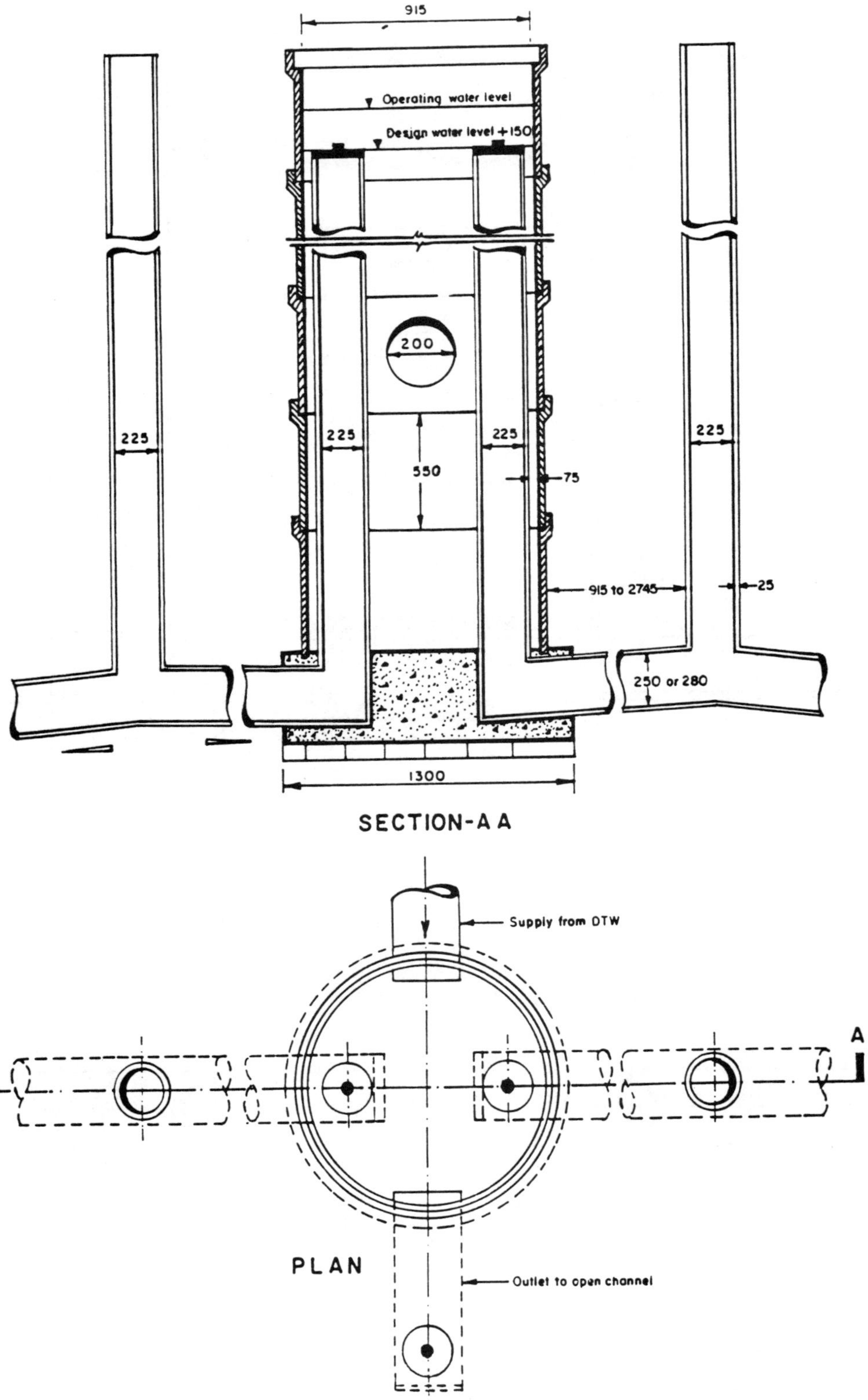

All dimensions in millimetres

Figure 5.4. Header tank with top control (using an internal pipe arrangement)

Source: MMI (1992)

One option is the use of downstream water-level-controlled Neyrtec gates, of either the AVIO or AVIS type as described by Goussard (1987). Such gates are installed in concrete housings, similar to those required for standard radial gates. The gates are self-operating and automatic, and therefore screening is required to prevent debris from entering the pipeline and interfering with closure of the gates. The gates should be attended on a regular basis. Maintenance will be limited to the greasing of bearings (on the large sizes of gates) and necessary painting.

Once set to the desired level the gate should remain within the design permissible range and not require readjustment. The pipe system itself will restrict the maximum flow, in both open and semi-closed systems.

Flow check and control structures

A wide range of structures is in use for the control of water flow in a buried pipe system. The structures differ in their operational flexibility and the degree of human intervention required. Most of the structures on pipe systems are variations on the open stand which is formed from either vertical sections of pipe or cast-in-place concrete or masonry. They may serve as gate stands or float valve stands with secondary functions as air vents, surge risers or sand traps. Design recommendations are provided for the following structures:

- overflow stands or checks;
- float valve stands;
- gate stands;
- sand trap.

ASAE standards (ASAE S261.7, 1989) recommend that:

- Wherever possible stands should avoid entrapping air in the pipeline by minimizing turbulence.
- They should be constructed with a minimum freeboard of 0.3m above the hydraulic grade line.
- Downward water velocities should always be less than pipeline velocities and preferably less than 0.6m/s, so the structures should be of larger diameter than the underground pipe.

Overflow stand or check

Overflow stands function as both checks and drops as well as acting as air vents and surge risers. As checks they set upstream water levels, so maintaining flow from hydrants and branch lines. As drop structures they reduce the pipeline operating pressures where required. Water falls either over a baffle or from a horizontal pipe connecting two vertical pipe risers.

Figure 5.5 shows an overflow stand which has been installed on closed systems in Bangladesh to prevent pipes from draining back towards the pump and into the well where no check valve has been provided.

On open pipe systems the height and spacing of overflow stands must be determined before design of the pipeline sections in between can be completed. Once the maximum height of the stand-pipe is known the spacing can be determined.

The height of the stand will be given by the hydraulic grade line. The maximum height should be less than both the following:

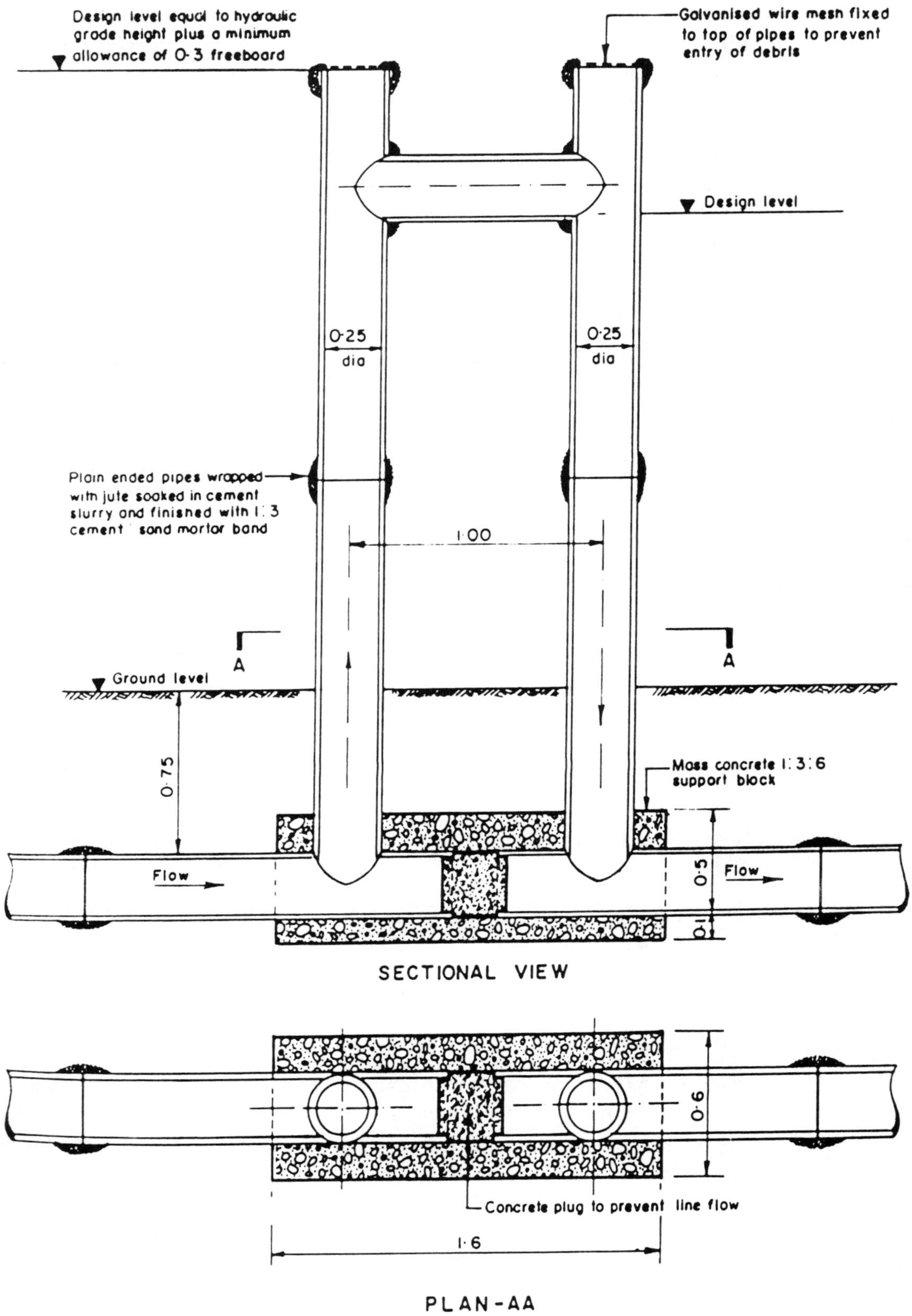

Figure 5.5. Buried pipe system overflow check stand

Source: MMI (1992)

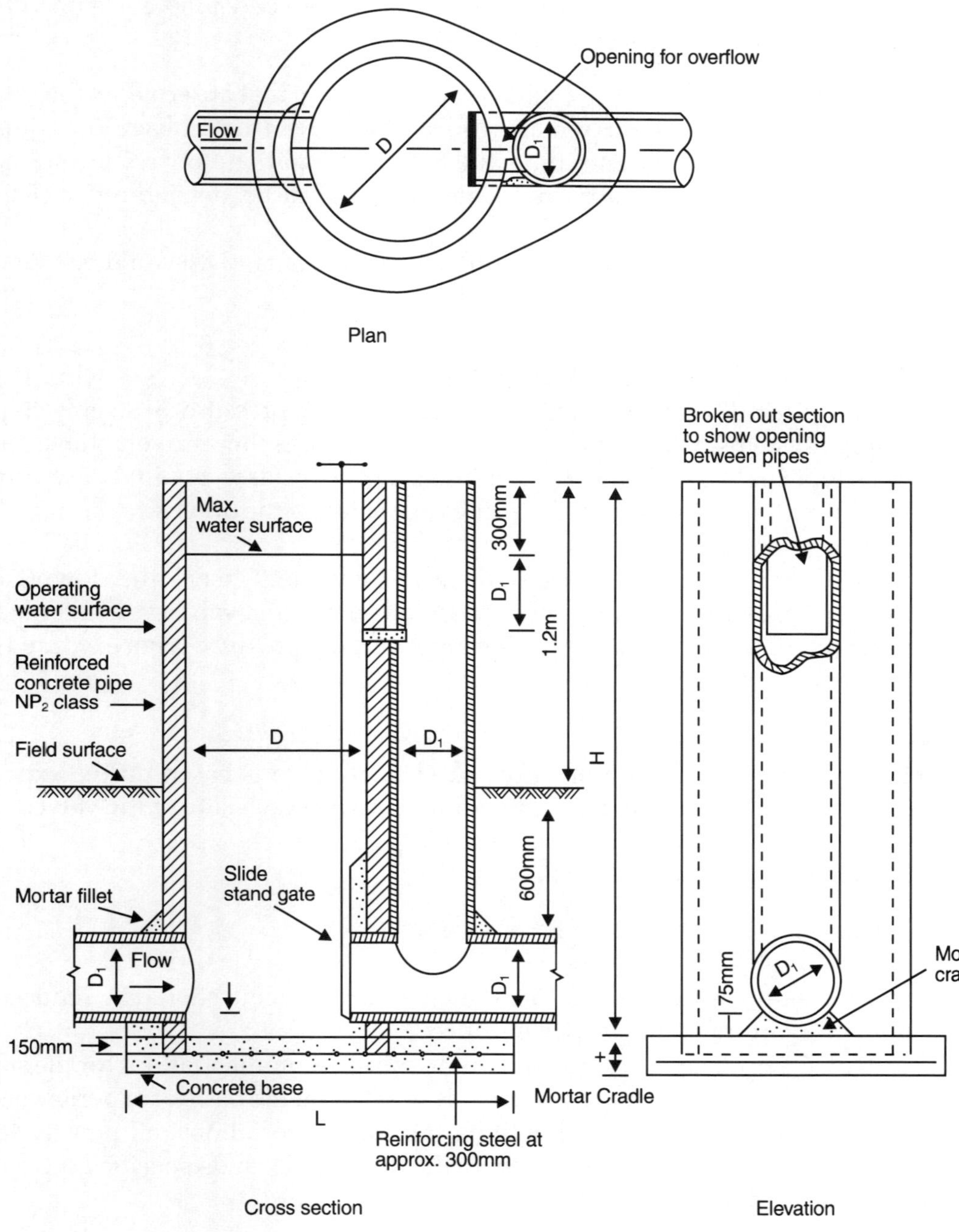

Figure 5.6. Overflow gate stand

Source: US SCS (1967)

- the maximum practical height to which the stand-pipe can be constructed: typical heights for non-reinforced concrete pipe stands will seldom exceed 3m above ground level;

- the maximum pressure rating of the pipe; Table 1.4 gives values for some of the commonly used low-pressure pipe materials.

The distance between the overflow stands should be such that at zero flow the section of pipe between the overflow stands remains full of water, given that leakage losses are not significant. This ensures that air entrapment, whenever irrigation flow resumes, is not a problem. Stand spacing will vary for a particular height of baffle, depending on changes in ground slope.

Figure 5.7 shows an isometric view of a concrete overflow stand as would be found on an open pipe system.

Float valve stands for semi-closed pipe systems

These structures function like the overflow stands to reduce pressure on steeply sloping semi-closed pipe systems. The float valves or Harris valves, as they are sometimes called after their developer, release into the stand only as much water as is required downstream with the float providing the water level sensing capability on the downstream side of the valve.

The height of the float controls the stand water level and hence the downstream pressure. Higher upstream pressures can be accepted with a float valve system compared with one using overflow stands as long as the pipe working pressure is not exceeded and as a result fewer structures may be required.

Selection

The selection of either a balanced or unbalanced float valve is based on the expected difference in pressure (Pd) at the float valve stand between each side of the valve.

Unbalanced float valve;	Pd < 8m
Balanced float valve;	8 < Pd < 15m

Where pressure differences exceed 15m then the stand spacing should be reduced, or an alternative pressure-reducing device installed.

For a particular size of valve there are manufacturers' recommendations for the size of float and the stand diameter. These have been developed on the basis of experience with a large number of installations and should be satisfactory for almost all pipe systems. Though uncommon there are situations where the floats and stands specified have been found inadequate, and these include:

- large diameter pipes used on long pipe reaches;

- where rapid flow rate changes are needed in the system and not just regulation;

- very long pipes on steeper slopes.

Different types of float valve (Merriam, 1987) are depicted in Figure 5.8. Cross-section and isometric views of a standard float valve are depicted in Figures 5.9 and 5.10 (US SCS, 1967). An external view of a float valve stand as constructed on a scheme in India is depicted in the publication 'Photographs of Buried Pipelines for Surface Irrigation'.

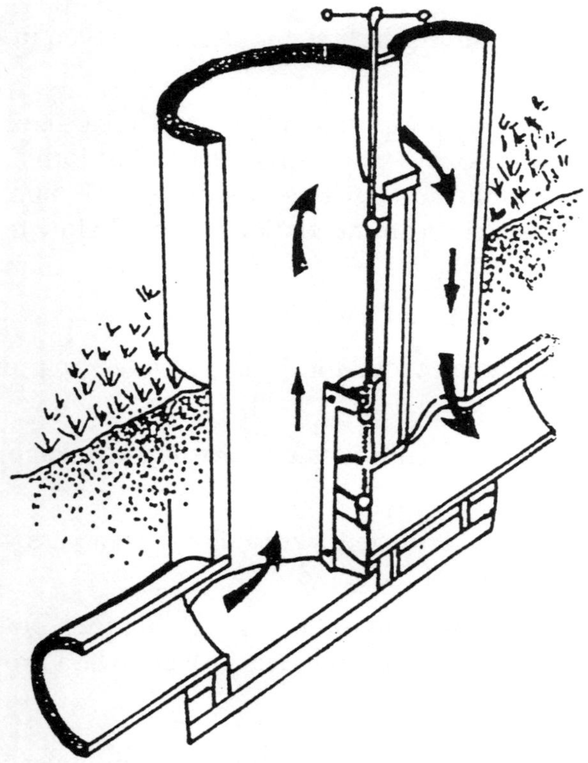

Figure 5.7. Overflow stand-pipe

Source: US SCS (1967)

Design

The energy lost through the float valve is the difference between the energy lost in pipe friction (sL) and the energy gained from the change in ground elevation (s_gL).

The design of float valves and their stands for semi-closed pipe systems is described in a step-wise procedure. In order to design the pipeline the following data must first be determined:

- ground slope of the pipe alignment (s_g);
- stand spacing which at its maximum will be determined by the maximum pressure rating of the pipe: the maximum pressure on the upstream side of the valve will be represented by the height of the hydraulic grade line at zero flow;
- minimum design energy head above ground (E): this sets the residual head in the float valve stand at design flow after valve head loss and any necessary throttling. This residual head will depend on the required minimum operating pressure at the outlet, immediately downstream of the stand.

In order to allow for even operation of the system the aim should be to maintain the minimum energy head at the design flow rate along the entire pipeline.

The design procedure summarized in Flow chart 4.2 is described by Merriam (1987), and involves the calculation of hydraulic conditions at design flow and full valve closure. Details including equations and definitions for the rotation used are detailed in Figure A3.1 in Appendix 3.

Head loss data for the float valves together with basic manufacturer's data is shown in Figure A3.2 also in Appendix 3.

A worked example of float valve selection and design is given in Appendix 3.

Construction

Operation of float valves with very few problems for over 25 years is reported by Chandler (1987) in the Orange Cove and Solano Irrigation Districts, California, USA. Vandalism is not reported, but this is a possible problem in some situations. The problems encountered will be avoided by following the recommendations set out below for constructing float valves and stands.

- The valve plates should be installed level.
- A float stem guide should be provided for the larger valves to prevent wearing of the floats against the stand walls.
- The styrofoam floats should be securely fastened to the rod linkage and a reserve clamp is a good safety measure.
- Like other stands it should be of sufficient size to allow easy access for maintenance with the valve installed on one side of the stand.
- Steel pins or steel sleeve bearings should be used rather then brass pins in the hinge joints of the valve. Rapid wear of brass pins has proved to be a problem in some cases.

Operation — valve hunting

Some problems may occur with the valve tending to over-adjust for flow changes or "hunt" when the valve is designed to prevent small fluctuations in water level. This tendency is often exaggerated for float valves in series (Merriam, 1987b). This can be

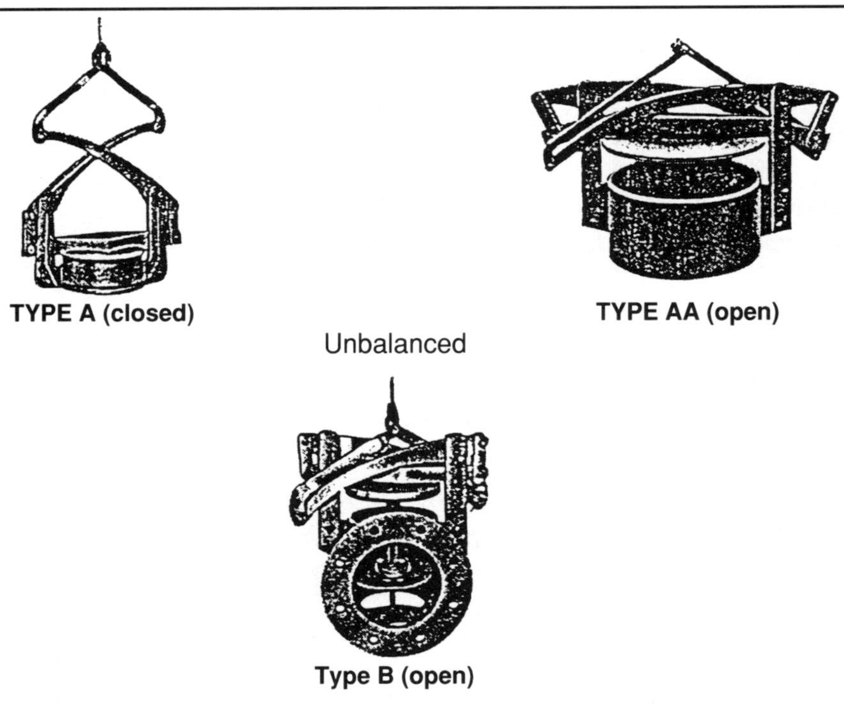

Figure 5.8. Types of Harris float valves

Source: Merriam (1987b)

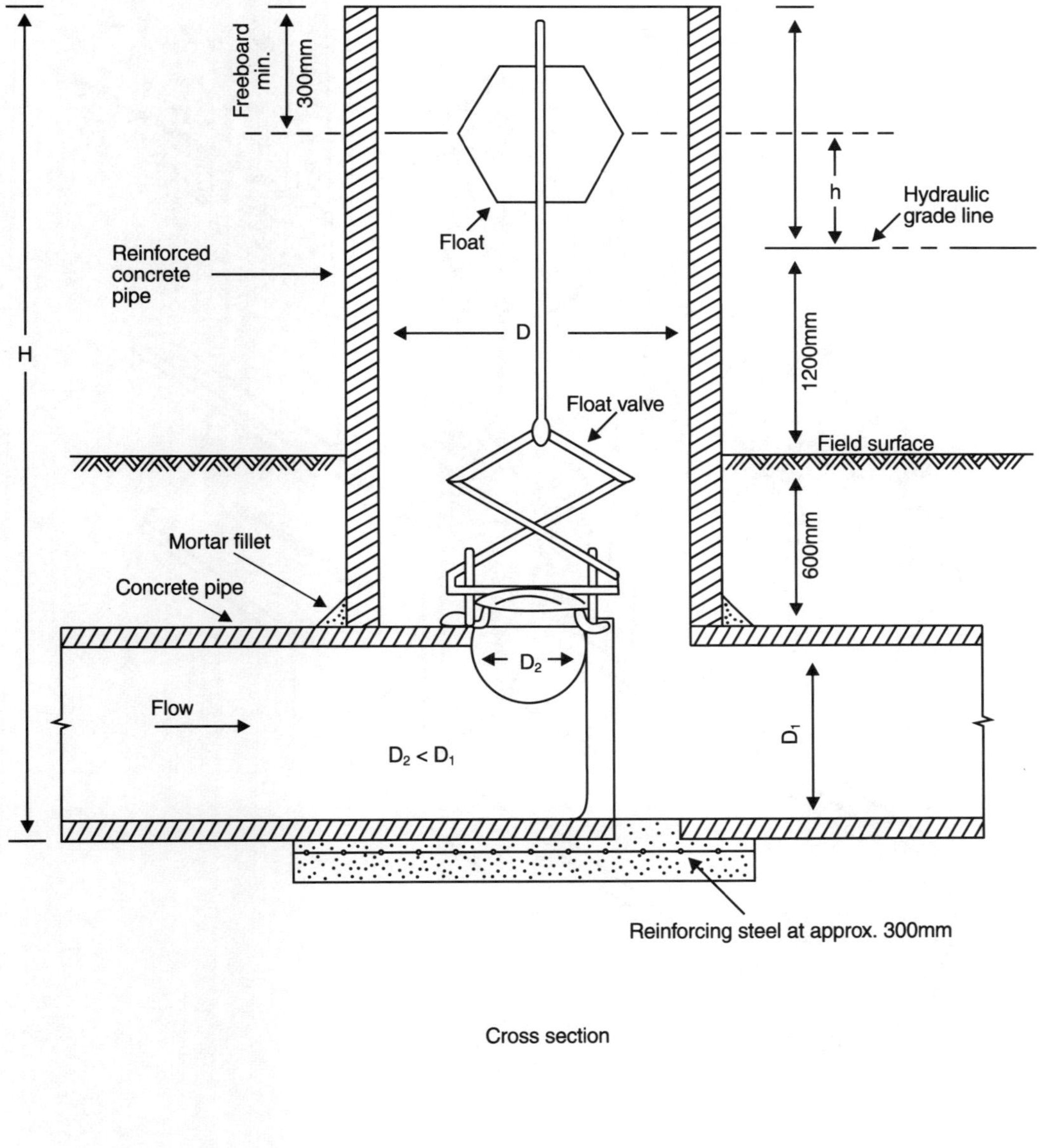

Figure 5.9. Cross-section of a float valve stand

Source: US soil conservation service (1967)

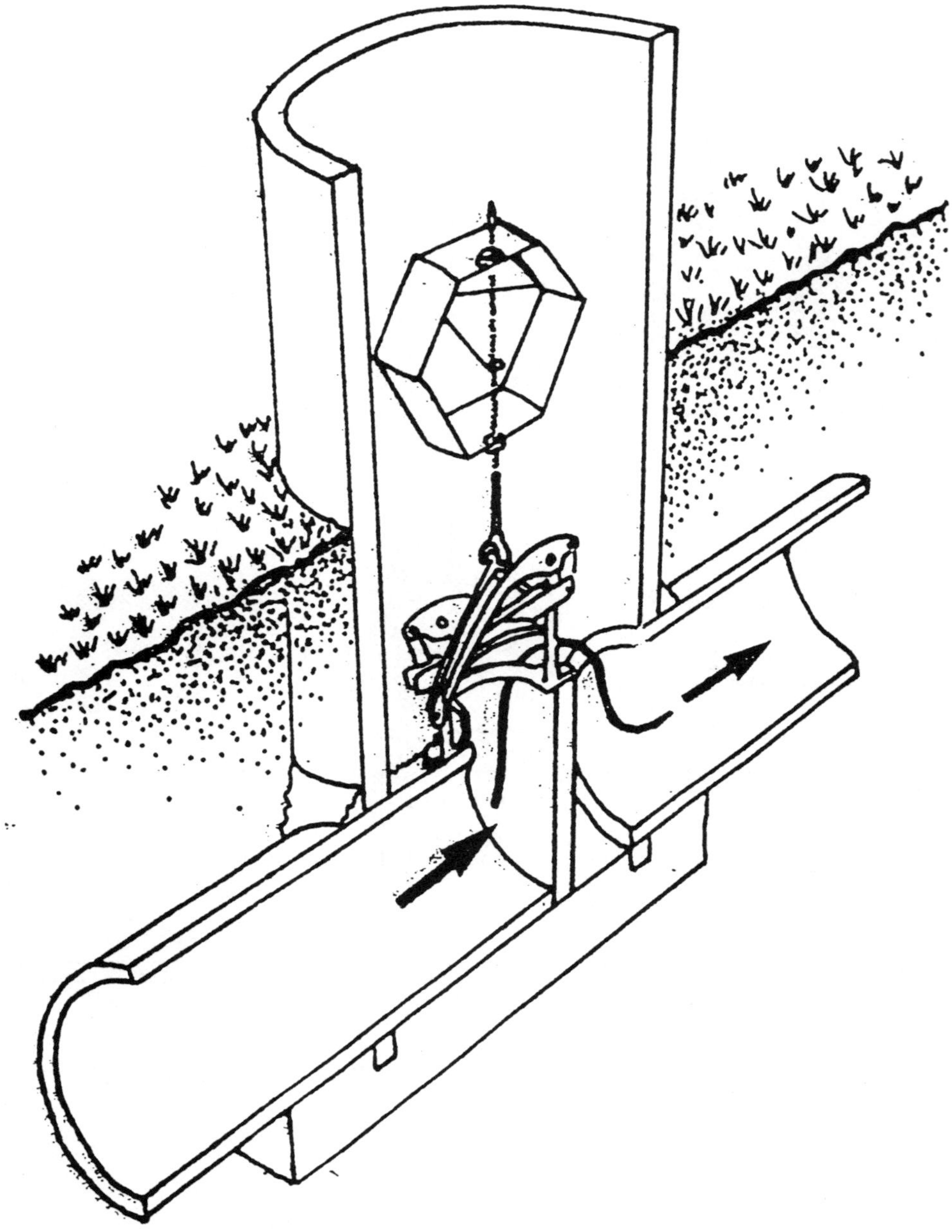

Figure 5.10. Section of a float valve stand

Source: US SCS (1967)

minimized by ensuring the change in water level between the float valve being fully open and closed is between 150 and 300mm. This can be assisted by using a thinner float, so that a greater change in water level is required for the same upthrust. Where "hunting" persists it can usually be controlled in one of two ways:

- The simplest adjustment is to reset the control height of floats in adjacent stands, one up and one down, so changing the elevation of the energy gradients.
- It is also practical to change the shape of alternate floats, while retaining the same volume. Shape change will alter the submergence and hence volume in the stands, so varying rates of valve closure.

Filling of the pipelines should be carried out slowly so that water can be released steadily without water hammer problems.

Pipelines should be kept full and where they are to be kept unused for some time a small amount of water for seepage can be admitted (on gravity-supply systems) by inserting a small pipe to bypass the closed shutoff gate (Merriam, 1987b).

Gate stands

Gate stands are mainly used to control the flow of water where a pipe system branches and supply needs to be directed to one or other branches. Header tanks and pressure towers function as gate stands where valves or gates control water entry to pipe branches and in most small pipe systems there should be no need for additional gate stands. The exception to this is where flow is to be divided between two branches in order to reduce the pipe size and/or friction losses. On larger pipe systems they may be required to control the division of flow but they need continual resetting when the selection of operating valves alters. They complicate operation by requiring manual setting.

On open pipe systems overflow stands are often fitted with gates to help reduce air entrainment, at peak flow.

Two arrangements of gates within the stand are common:

- Where pipes enter the side of the stand, slide gates are used. These are usually opened by means of a threaded gate stem.

  The use of a removable wheel to operate the gate discourages any tampering with gate settings. A gate stand of this type is shown in Figure 5.11. Direct acting slide gates should not be used except on the very smallest systems as they become difficult to operate with any cementing or corrosion of the gate.

- The stand is built above the pipes which have openings in the floor of the stand. Alfalfa valves fitted into the floor of the stand are opened and closed by the use of long handles which key into the valve lid.

Guide frames should be installed to keep the handles in place, as it becomes difficult to position the handles with water in the stand. Such frames have been used for the header tanks built on systems in Bangladesh under TADP (Georgi, 1989) and IDA DTW II Projects (MMP, 1989a). The handle arrangement and guide frame are detailed in Figure 5.12. These features are illustrated in the publication 'Photographs of Buried Pipelines for Surface Irrigation'.

Gate stands are usually made from large diameter concrete pipe either precast or cast in place. In some cases stands have been made from stone or brick masonry. They should be of diameters that allow easy access for the repair of the gate or valve mechanism with a minimum diameter of 0.75m (Koluvek, 1970). Situations in which special gate stands, apart from the header tank, are provided in a pipe system, are described in a number of texts for USA-built systems, for example by Pillsbury and by Pimley and Fischer (1990).

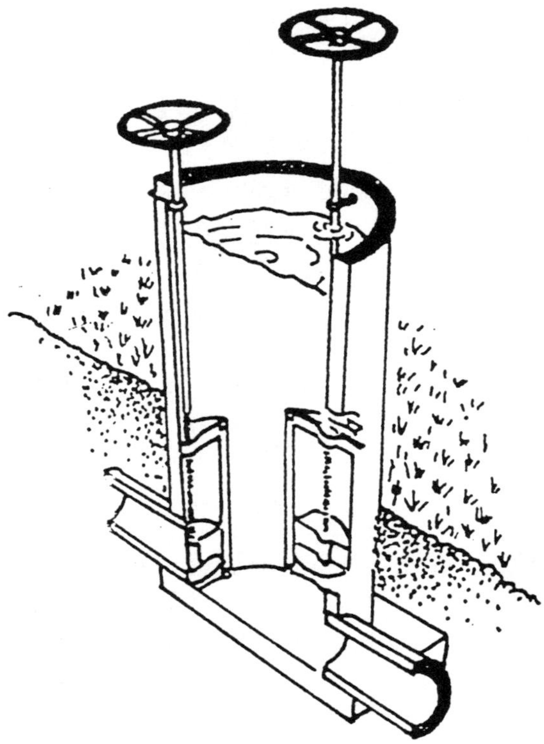

Figure 5.11. Section of a gate stand

Source: US SCS (1967)

Sand traps

Sand traps will normally be located between the water source and the pipe system and are only practical where the stand is open to the atmosphere. Sand traps may be necessary as part of the inlet structure in a gravity-supply scheme or be incorporated into pumpstands where water is drawn from a surface source high in solids. They should not be necessary for groundwater-supplied systems where large quantities of solids usually indicate inadequate well construction or screening and will increase the rate of pump wear. A typical design of sand trap is shown in Figure 5.13.

The sand trap will remove only larger sand-sized particles, as silt and clay material will not settle at the flow velocities normally encountered through the sand trap. In most cases where silt or clay particles are able to settle within the buried pipe systems, velocities in the pipeline when the nearby outlet is opened (0.4 to 1.5m/s) should be adequate to scour any deposits.

The dimensions of a sand trap should relate to the expected volume of material settling and a conservative estimate for the frequency of cleaning.

Construction

Koluvek (1970) makes the following recommendations:

- the inlet and outlet pipes to the stand should be vertically offset by at least two times the pipeline diameter;

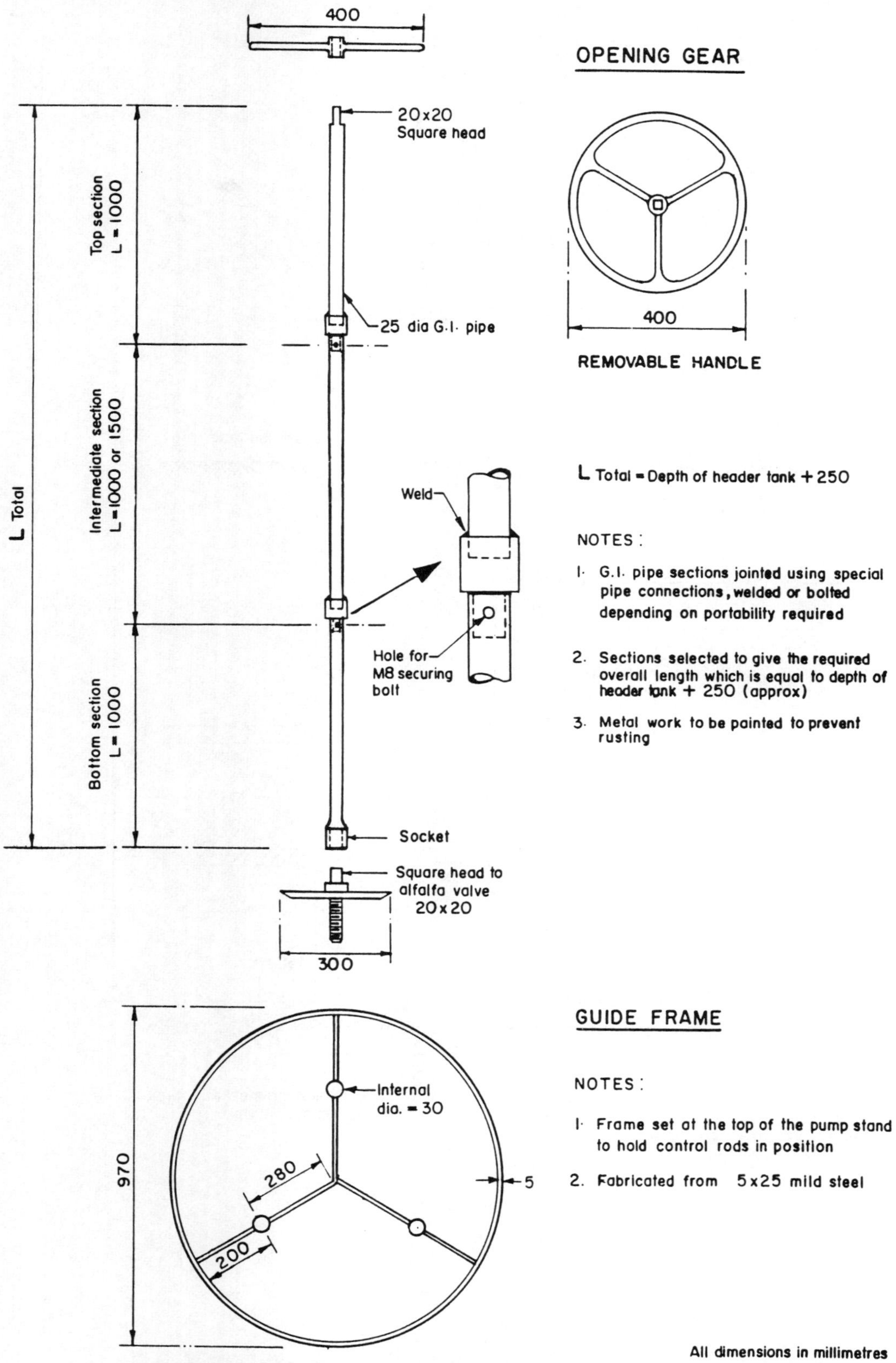

Figure 5.12. Header tank and division control mechanism

Source: MMI, 1992

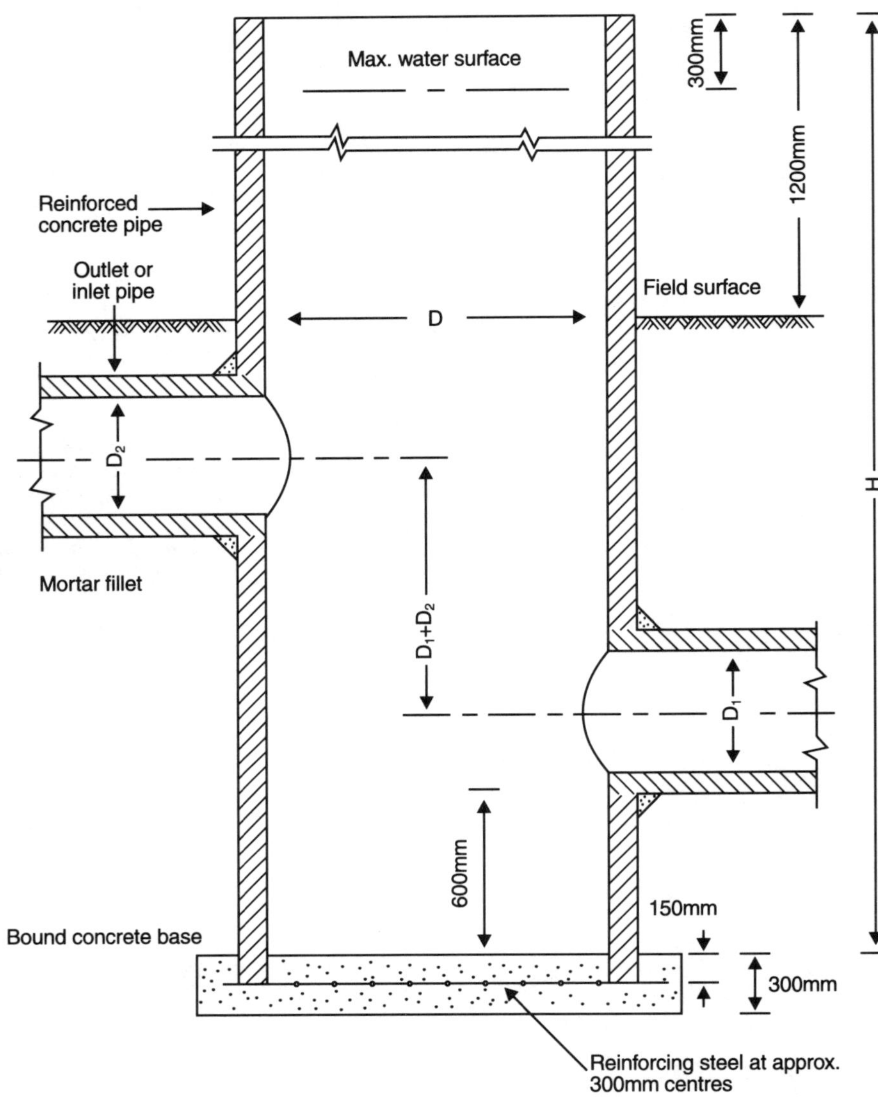

Figure 5.13. Cross-section of a sand trap

Source: US SCS (1967)

- pumpstands serving as sand traps should have a minimum inside diameter of 0.8m and a minimum depth below the invert of the outlet pipe of 0.6m;
- the stand should be of adequate size such that the vertical flow velocity does not exceed 0.1m/s;
- a pipe of minimum diameter 100mm should be provided to facilitate sand removal on sloping sites.

Surge risers

In practice any open pipe stands, including the header tank or pressure tower as well as those specified for flow control or as overflow stands, will act as surge risers.

The top of the surge riser must be above the hydraulic grade line for operation of any outlet in the system. This should include an allowance for freeboard (0.3m) to avoid spilling from small head fluctuations during operation.

Campbell (1986) states the following requirements for surge risers if they are to be effective in reflecting water hammer:

- The riser must be connected either directly to the line or via a short up-slope branch of less than 30m.
- The diameter of the riser should not be less than that of the pipe to which it is attached.
- The riser should not be less than 200mm in diameter where minimizing negative pressure is important, for example on thin-walled uPVC pipes.

The maximum height of the riser above the pipe must not exceed the safe working pressure of the pipe and will also be limited by the construction methods and materials used. In common with air vents, the maximum practical height for a riser built using uPVC pipe will be about 5m while for non-reinforced concrete it would be only around 4m.

Like air vents, surge risers, being above-ground structures, are susceptible to vandalism and damage. Similar methods to those mentioned for air vents can be used, to help reduce maintenance and repair costs. A typical surge riser as built by TADP on systems in Bangladesh, which also functions as an air vent, is shown in Figure 5.14.

Air vents and valves

Air vents are installed primarily to allow for the slow and controlled release of air. As they are above-ground structures, only the minimum number of air vents should be provided to keep the cost of maintenance to a minimum.

Air is found in all pipelines and if not vented in a controlled way will result in surge, which has already been alluded to, as well as a reduction in the flow capacity of the pipeline. Air enters the pipeline in a number of ways:

- the vortex action of the pump;
- through entry into the suction end of a pipe;
- when the pipe is empty;
- in solution;
- from entrainment at the water supply inlet.

Although the situations where air vents are likely to be necessary can be predicted, it is difficult to anticipate all situations without erring on the side of over-design. In most pipe systems where a problem occurs after commissioning it is a relatively simple

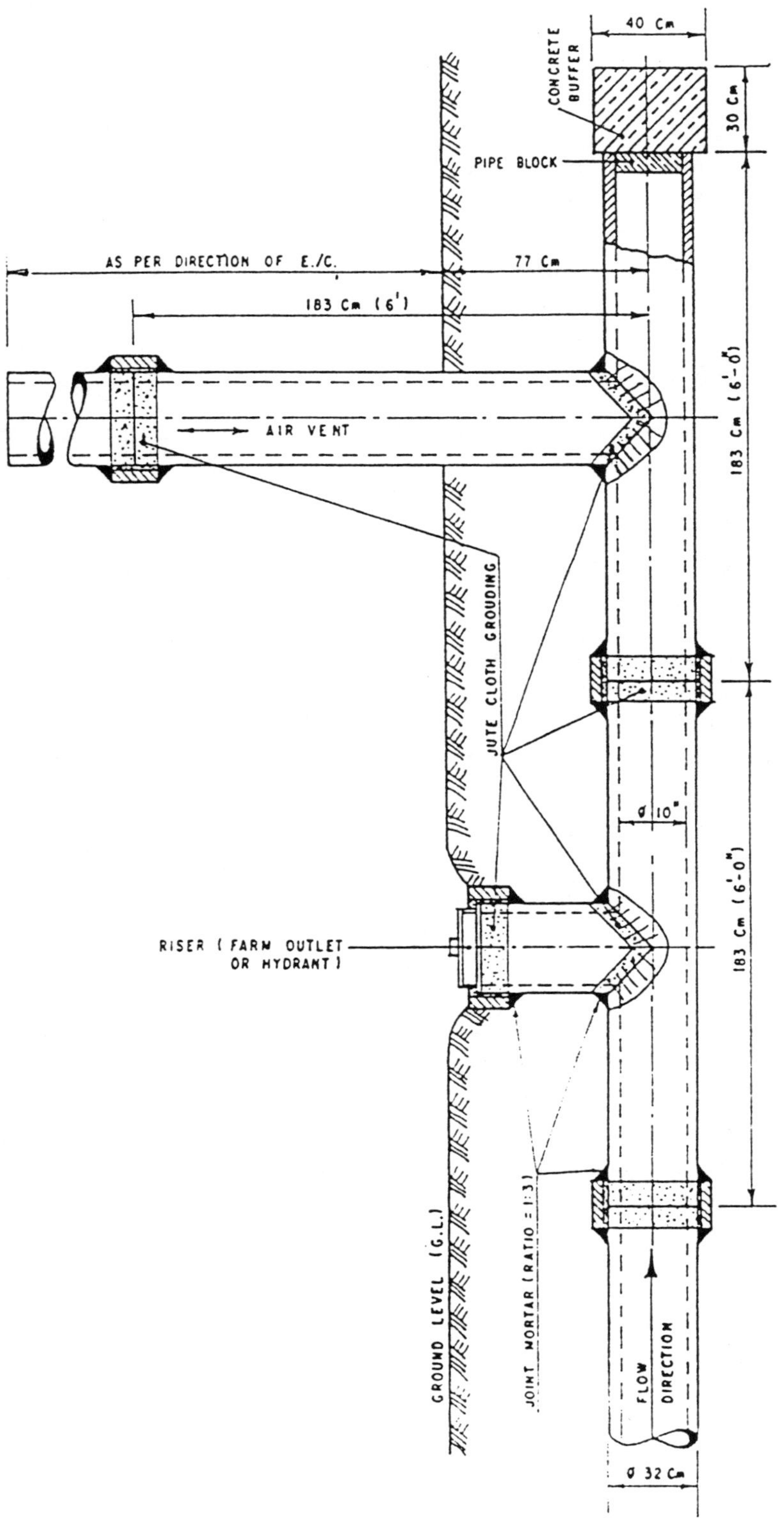

Note: The concrete buffer should be anchored
1 foot (30cm) deep in both site of the trench
in to the undisturbed soil, total length = about 5 feet.

Figure 5.14. Air vent/surge riser (Bangladesh model)

Source: Georgi (1989)

operation to fit an additional air vent later. Standards provide some guidance on spacings for air vents and suggest a minimum spacing of 400m and preferably 200m (ASAE S261.7, 1989) but experience gained with design and operation at site will provide the most reliable guidance.

Situations where air vents are particularly necessary include:

- very flat pipeline systems where the movement of air is slow during filling;
- where the pipeline empties to a substantial extent between irrigation events, for example, non-reinforced concrete pipe with mortar joints;
- at positions where substantial quantities of air become entrained in the flow:
  - i) at overflow stands in open pipe systems;
  - ii) in narrow pump stands with high velocities through the stand;
- sharp changes in pipeline alignment, particularly at high points and downward changes in grade in the direction of flow;
- downstream of in-line shutoff valves for vacuum relief, for pipe materials susceptible to negative pressures.

Branching pipe systems will normally require more air vents in the same circumstances than loop systems which fill from two directions. Wherever possible, systems should be designed to minimize the amount of air entrained or the degree to which the pipe system empties between irrigation events.

Structures to vent air fall into two groups: vents and valves.

Vents

These comprise open stand structures which are used on low-pressure pipelines, as the height of the vent must extend above the hydraulic gradeline. These structures are considered in more detail below under the heading "Open vents". An alternative vent structure is a mechanical vent which operates only to release large volumes of air.

Valves

Allow for the venting of air at higher pressures where open vents are not practical. Sizing of the orifice of an air release valve will be based on the pipeline diameter, pipe slope, flow rate and operating pressures. Where pipeline filling occurs regularly, for example where seepage losses cause pipeline emptying between irrigation events, large-orifice air release valves need to be provided. These are poor at venting small amounts of air during normal operation and additional continuous-acting air valves often need to be provided.

Vacuum valves are provided to allow air entry to prevent a vacuum forming which can cause the collapse of pipes like thin-walled uPVC.

Open vents

These are vents connected directly to the pipeline and extending to a height above the pipe invert equal to the hydraulic grade line plus a minimum freeboard of 0.3m. In general they vents are not used for heights above 4 to 5m. Where the vent is a minimum of 200mm diameter it will also function as a surge riser in reflecting pressure waves due to water hammer or surge. The American standard ASAE S261.7 (1989) states that where surge protection is not necessary the large diameter vent can be replaced by a vent 1/60 of the cross-sectional diameter of the pipeline but not less than 50mm. This size reduction can begin at a height not less than one pipeline diameter above the pipe centreline. The pipe should protrude inside the vent (about 50mm) so that the small volume of air trapped provides a surge cushion.

Table 5.1 **Minimum orifice size for large-volume air release valves**

Pipe diameter (mm)	*Min orifice diameter (mm)*
150	50
250	55
380	80
500	100
760	125
910	165

Source: ASAE Standard S261.7

Two alternative air vent structures are shown. The first (Figure 5.14) is typical of the open vents used on closed pipe systems in Bangladesh. The second design (Figure 5.15) uses a smaller diameter steel pipe which could be replaced with a mechanical air release valve, of the type described below. This type has been used on loop uPVC pipe systems built in Madhya Pradesh, India. Examples of other commonly used open vents are illustrated in the publication 'photographs of Buried Pipelines for Surface Irrigation'.

Mechanical air vents

These valves can be used as alternatives to open vents on low-pressure pipelines. They allow for the release of large volumes of air, particularly important during pipeline filling, while remaining closed when the pipeline is full of water during normal operation. They usually extend less than 0.5m above the ground. These valves tend not to release small amounts of trapped air during pipeline operation.

Recommendations on minimum sizes of these valves are given by the American standard ASAE S261.7 (1989) and are summarized above. Apart from references to the manufacture and use of large concrete valve chambers with an internal ball float on Chinese buried hose systems (Chengzhi, 1988), manufacturers advertise the availability of proprietary valves of this type which only seal well between pressures of 7 and 35m. Refer to Figure 5.16 (Waterman Industries, 1990).

Vacuum relief valves

These valves are supplied as either single-acting, vacuum release valves or dual-acting, where they also function as air release valves. Vacuum valves are particularly important

Table 5.2 **Minimum air inlet orifice diameters — vacuum valves**

Pipe diameter (mm)	*150*	*250*	*380*	*500*	*760*
Percentage slope	*Minimum air inlet orifice diameter (mm)*				
1	15	55	90	115	175
2	25	60	100	130	200
3	30	70	105	145	220
4	40	75	110	155	235
6	50	85	120	160	260
10	55	90	135	175	315
20	60	100	160	210	370

Source: ASAE Standard S261.7

on thin-walled uPVC systems which are more susceptible to collapse due to negative pressures.

A vacuum can occur as the result of a break in the pipeline which causes the pipe to drain rapidly. Alternatively during normal operation a vacuum can be established if the hydraulic gradient is below the top of the pipe or it can occur when a pipeline is drained during routine maintenance. A vacuum valve can also be used as a siphon breaker. The requirement for vacuum release valves should be kept to a minimum by careful design.

Particularly important positions for air/vacuum release valves are downstream of pumps connected directly to the pipe system and rapid closing in-line gates or valves. Low-pressure concrete pipe systems do not usually require these valves where open vents are used.

High pressure pipe systems require the installation of mechanically-operated air release and air/vacuum valves, which are actuated at specified trigger pressures.

A number of manufacturers offer proprietary valves which function well at medium to high pressures (see Figure 5.17). The reliability of their performance at lower pressures is unclear. Recommendations for the minimum orifice sizes for vacuum relief valves installed at high points, as detailed in ASAE S261.7, are shown opposite.

Continuously acting air release valves

These valves cannot cope with the release of large quantities of air but are practical where air becomes trapped in small quantities during continuous operation. They can be used in pipeline systems which do not empty to any great extent between irrigation events or in addition to large orifice valves.

Manufacturers advertise proprietary valves which combine vacuum and air release and operate from pressures as low as 3m (see Figure 5.18). Recommendations on the orifice size of the valves, based on the pipeline flow rate, are shown below, as detailed by ASAE S261.7 (1989).

Construction

Air vents

The positioning of air vents is best done at the same time as system design for surge risers, given that all surge risers can act as air vents. Surge risers can be repositioned for the purposes of air venting or air release in some cases, without significantly affecting their performance. For purposes of standardization extra air vents may be constructed from the same material as the pipeline although vents using galvanized iron pipes of minimum diameter 50mm are an alternative.

Table 5.3 **Minimum orifice size for continuously acting air release valves at different pipeline flow rates**

Flow rate (l/s)	*Orifice size (mm)*
0-25	2.5
25-50	4
50-100	5
100-150	6
> 150	10

Source: ASAE Standard S261.7

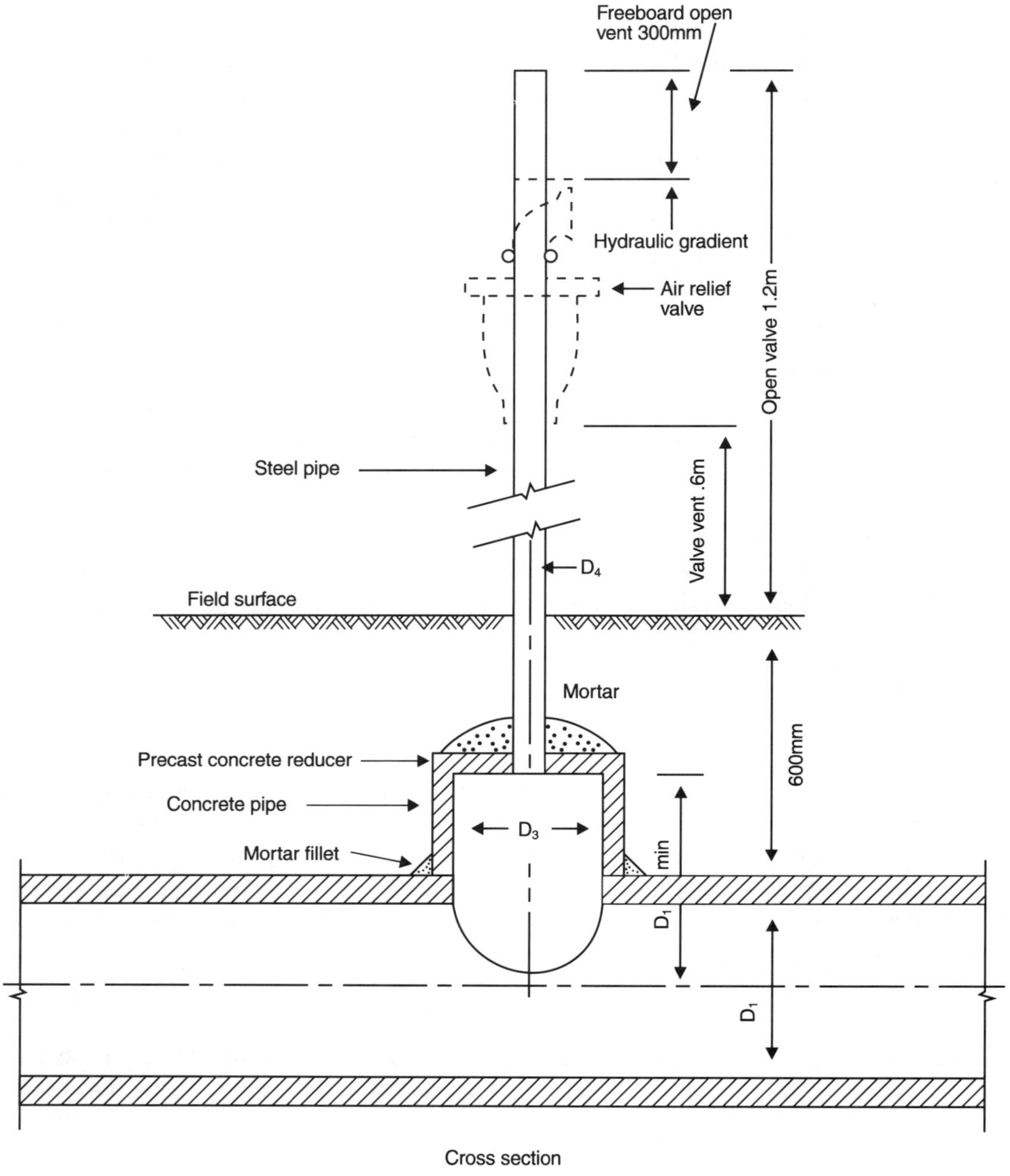

Figure 5.15. Air vent

Source: US SCS (1967)

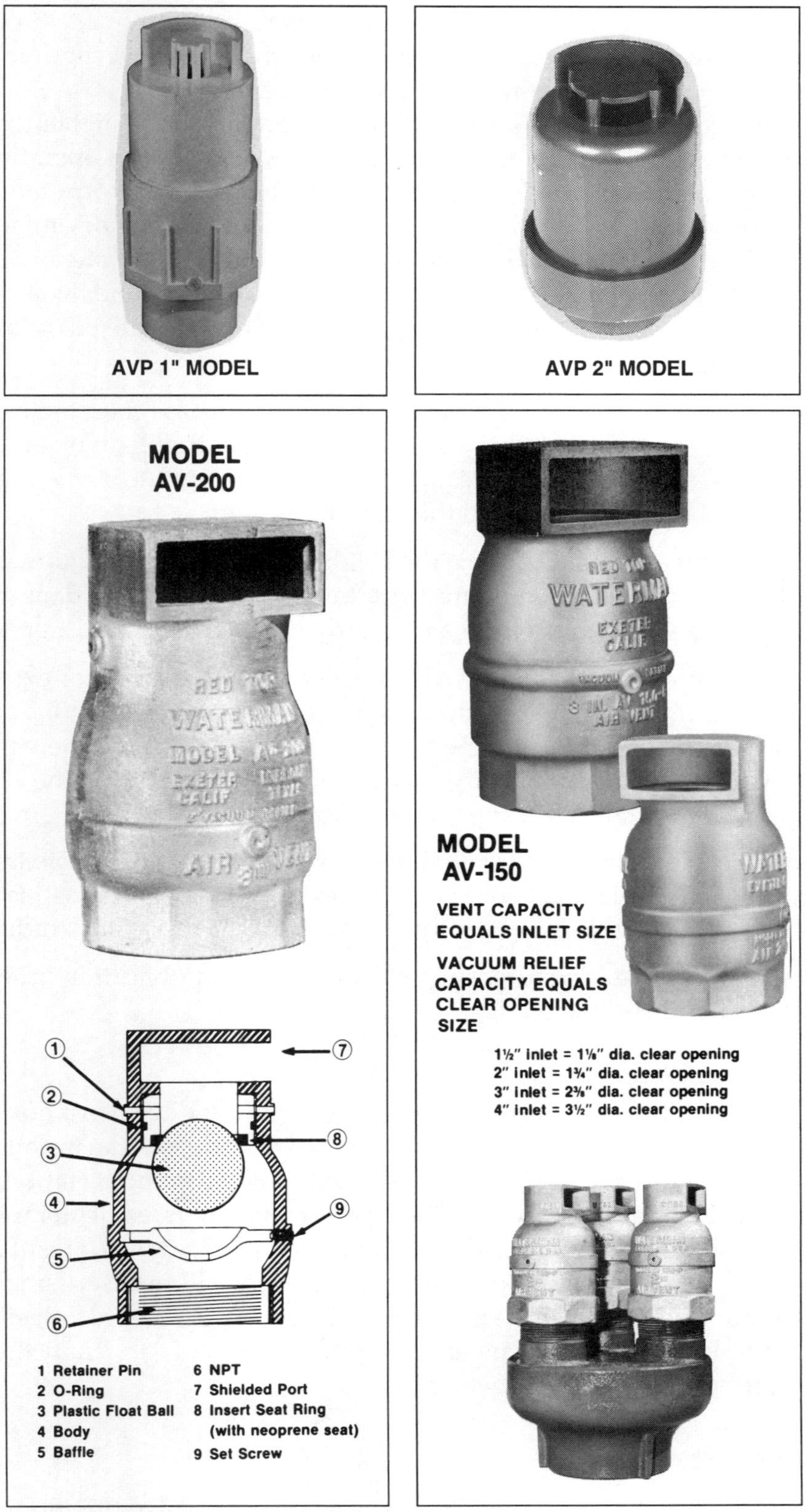

Figure 5.16. Non-continuous mechanical air vents

Source: Waterman industries (1990)

The maximum practical height to which an air vent can be constructed depends on the material used. For non-reinforced concrete pipe this is about 3m above ground level while for uPVC pipe up to 4m is possible. The required height of the air vent (hydraulic grade line + freeboard) must be determined for each open vent.

Low-pressure concrete pipe systems will seldom require vents higher than 3m, however uPVC pipe systems may require air release at positions with operating pressures beyond the scope of open vents. Mechanical air valves, which perform more reliably at medium pressures, are the best choice in these cases, and are an essential component downstream of the pump on any direct connected pump supply system.

As above-ground structures, air vents are susceptible to vandalism, and various strategies have been employed to tackle the problem. Of these, strategies which should be considered include:

- Explain to the system users of the function of the structure as ignorance can encourage vandalism and even pilfering of above-ground materials. Problems of removal of surge risers on uPVC pipe systems in Uttar Pradesh were reduced by improving the explanation of the system design to the farmers.
- Ensure that open vents extend above the height which would provide easy access for children. For most air vents pipe diameters are too small to present a danger to children. Larger surge risers and stands for water control are likely to prove a greater hazard.
- On open vents provide a perforated cap or covering with a screen or expanded metal grate to prevent the entry of large diameter objects (for example, stones and clods of earth) which could block the pipeline. This may not be necessary where the air vent is particularly high. On the uPVC surge risers and air vents used on the Uttar Pradesh uPVC pipe systems, the riser is topped with a perforated cap.
- Provide a U-bend on the top of small diameter vent pipes to prevent material entering the pipe. This technique has been used on the buried pipe systems installed under USAID-funded pilot projects in Madhya Pradesh and Maharashtra, India.

Examples of the above structures are illustrated in the publication 'photographs of Buried Pipelines for Surface Irrigation'.

Air valves

Air release and vacuum valves can be installed below ground level in a box surround or they may be installed on a riser above ground level. In both cases the installation should include a shutoff valve on the pipe side to allow their removal without emptying the entire pipeline. Access for maintenance should be easy and, if freezing temperatures are expected, protection should be provided; this may include lagging or insulation.

These valves, and in particular the air release valves, will release a small amount of water under normal operation and larger quantities if foreign material lodges in the valve, preventing complete closing. Arrangements should be made to drain the water and prevent a waterlogged area from developing.

Outlets and risers

An outlet is a device to deliver water from a buried pipe system to the surface for further distribution to the field. In the widest sense an outlet consists of a valve, riser pipe and/ or an outlet gate.

The choice of outlet will depend largely on the method of surface distribution of the water to be used and the maximum flow rate envisaged at the outlet. However, by far the majority of buried pipe systems will be adequately served by a standard alfalfa valve or a locally manufactured equivalent. The typical riser and alfalfa outlet, discharging directly into an earth channel, are shown in Figure 5.19. The selection of an outlet valve is summarized in Flow chart 5.1.

Where field distribution involves the use of surface pipe or hoses, one of the following may be appropriate:

- portable hydrant (see Figure 5.20);
- surface pipe outlet (see Figure 5.20);
- swivel arm distributor;
- fixed hydrant.

Where distribution involves earth channels the choice is on the basis of whether the flow is to be split at the outlet (rather than in the distribution structure). Outlets which divide flow through gates or openings into several streams include:

- gated outlet;
- capped riser or pot (see Figure 5.20);
- open pot (see Figure 5.21);

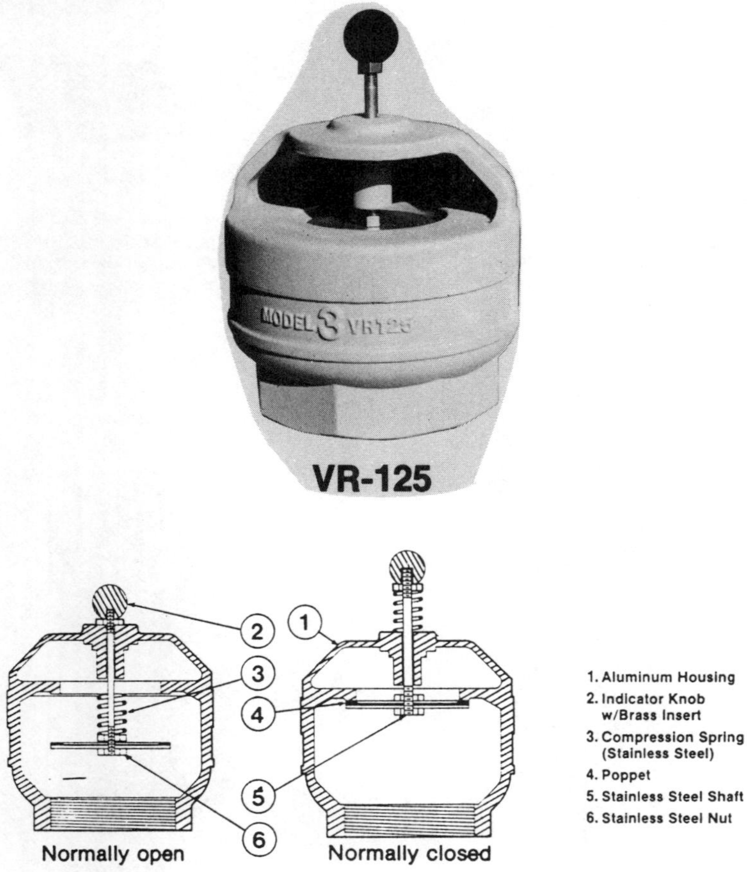

Figure 5.17. Air and vacuum relief valves

Source: Waterman Industries (1990)

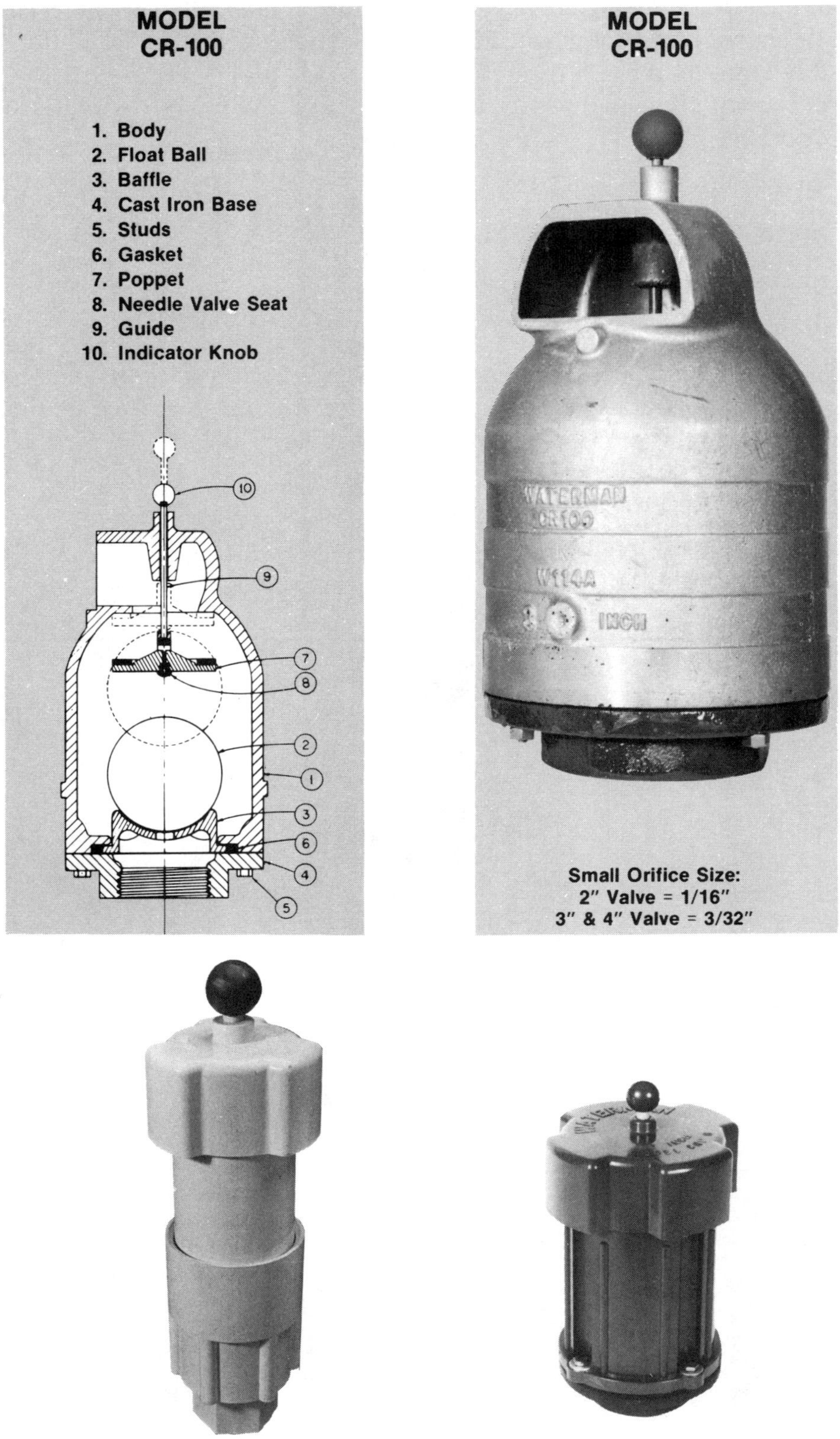

1" Model **2" Model**

Figure 5.18. Continuous air release valves

Source: Waterman Industries (1990)

- open stand with side orifices.

Where the full outlet flow is delivered to the earth channel or distribution structure, alternative outlet choices include:

- alfalfa valve (see Figure 5.21);
- modified alfalfa valve (US SCS, 1967);
- orchard valve (see Figure 5.21);
- end capped riser.

The modified alfalfa valve allows for the attachment of a portable hydrant to supply a pipe or hose. Orchard valves are designed for flows smaller than those typical through alfalfa valves.

Examples of different outlet valves in use are illustrated in the publication 'Photographs of Buried Pipelines for Surface Irrigation'.

Alfalfa valve

This comprises a screw valve, usually attached by cement to the top of a pipe riser. The valve opening is equal in diameter to the inside diameter of the riser pipe. The valve lid is attached to a threaded rod or spindle which moves up and down as the lid is moved by hand or by means of a handle or key inserted into the lid. When it is opened, water flows from all sides of the valve. A typical alfalfa valve as manufactured in Bangladesh is shown in Figure 5.24.

The most important aspect of design is selection of a valve and riser of adequate diameter for the maximum flow through the outlet. The head losses through alfalfa valves depend largely on the flow velocity but also to a limited extent on the shape and design characteristics of the valve itself. Typically the head loss will be in the range of 1.2 to 1.5 times the velocity head:

$$\frac{V^2}{2g}$$

The choice of valve size should be such that the head loss is kept below 0.15m.

Friction loss estimates for alfalfa valves of different diameters are detailed in Table 4.3 in Chapter 4 (Booher, 1974). Based on these estimates and allowing for a maximum head loss of 0.15m under low head and 0.65m under high head, recommended maximum flow rates have been derived for different alfalfa valve sizes and are detailed in Table 5.4. More complete alfalfa valve head-loss data are presented in Table 4.3.

Alfalfa valves in use are commonly lower-cost, locally manufactured versions of the standard alfalfa valve from the USA. The head-loss characteristics of these valves differ from those quoted by valve manufacturers though the extent depends on the modifications which have been made to simplify fabrication.

Design and construction

Because the valve is one of the few moving parts on the pipe system, careful specification of its design and construction can help reduce the frequency of damage and replacement.

Design recommendations

- Because the valve will inevitably require repair and replacement, the materials and manufacturing methods used should be as inexpensive as possible.

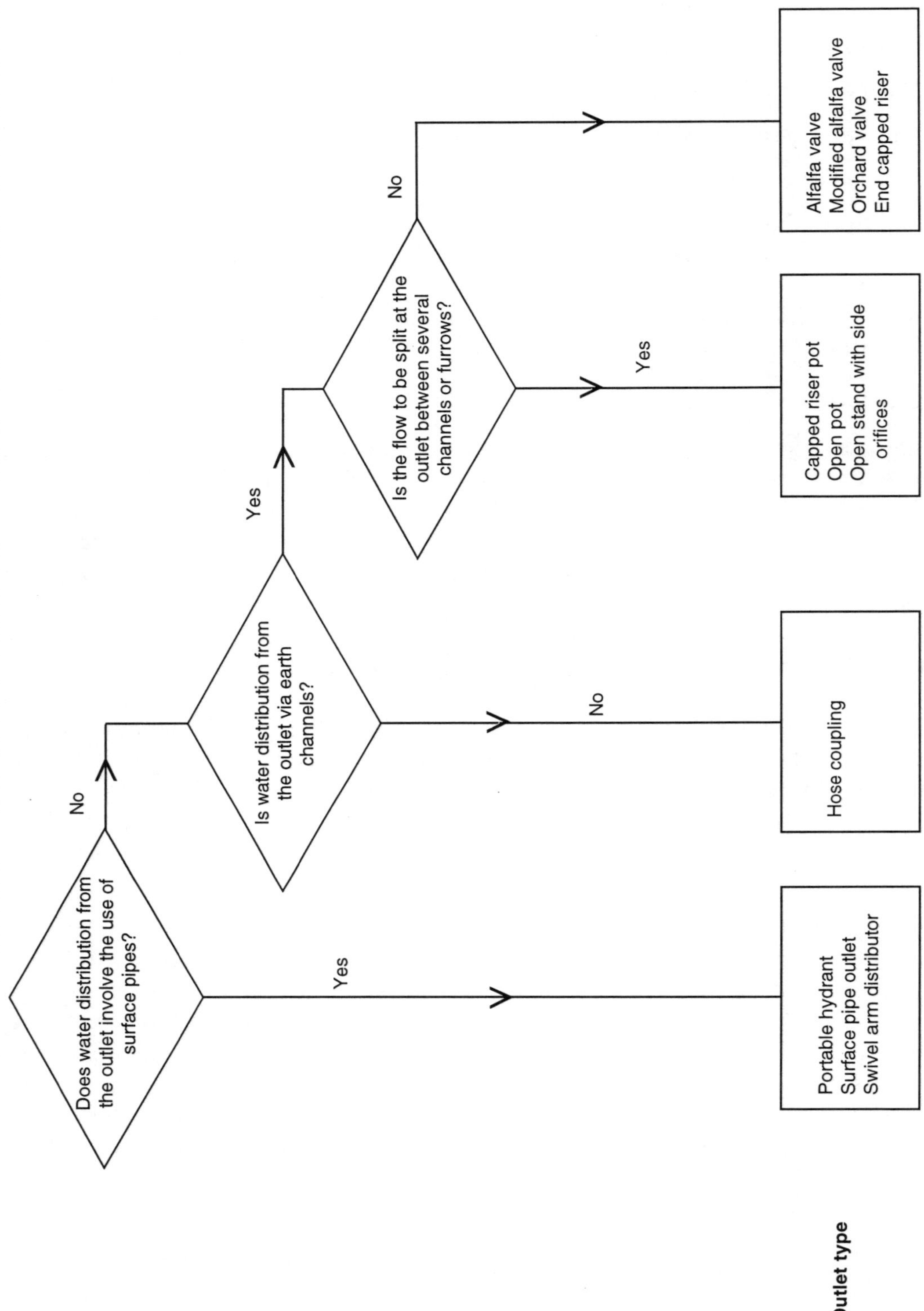

Flow chart 5.1. Selection of outlet type

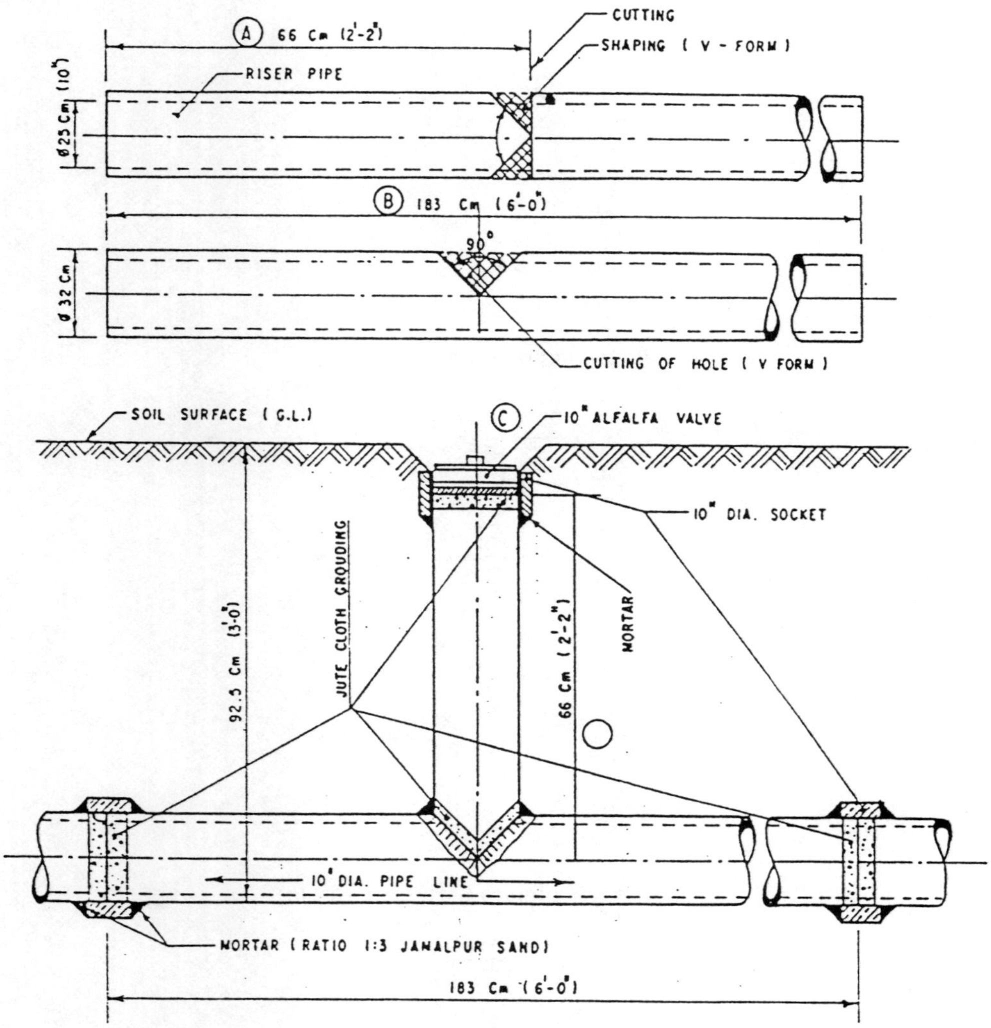

Note:

1) The seat of the lid of alfalfa valve has to be minimum ¼" above socket
2) Top of lid (alfalfa valve) C, 3" below ground level (G.L.)
3) Length of pipes related to spanned pipes

System \ Measurement	A (cm)	B (cm)	C (in)
8"	72 cm	100 - 110	8"
9"	68 cm	100 - 110	8"
12"	60 cm	100 - 110	10"

Figure 5.19. Riser valve design

Source: TADP. Georgi (1989)

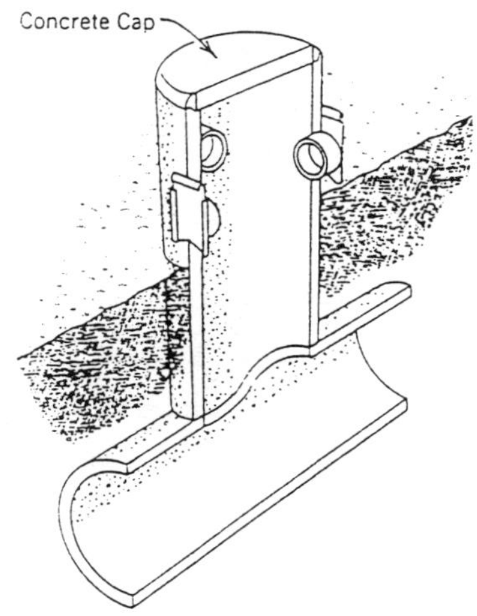

i) Capped risers or pot outlet

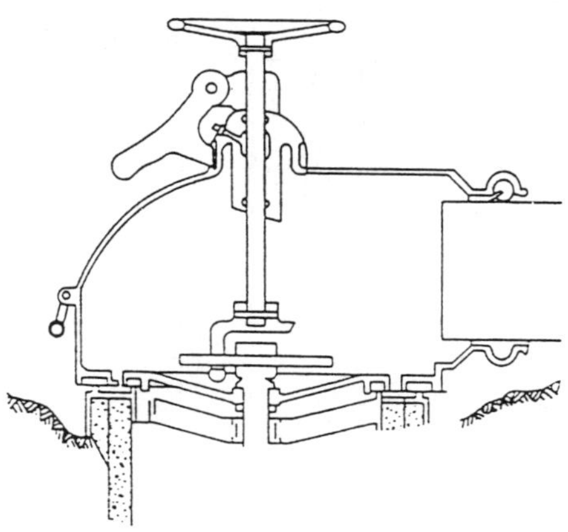

ii) Cast aluminium portable hydrant

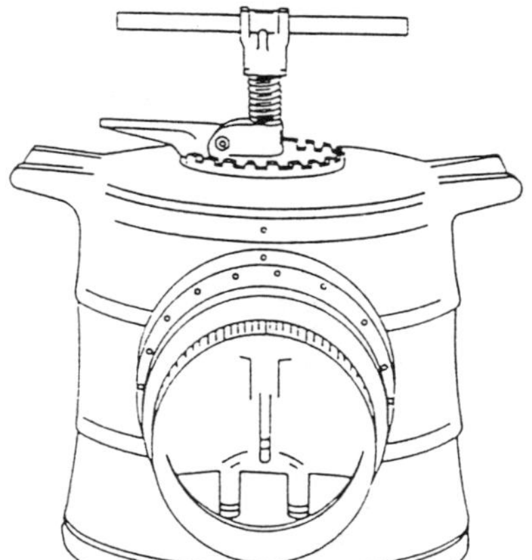

iii) Sheet metal portable hydrant

Figure 5.20. Alternative outlets 1

Source: James (1988)

Table 5.4 **Alfalfa valve maximum design capacities**

Inside dia. of riser (mm)	*Dia. of port (mm)*	*Maximum design capacity (l/sec)*	
		Low head (0.15 m)	*High head (0.65 m)*
150	150	22.6	45.3
200	200	39.6	79.3
250	250	62.3	124.5
300	304	87.7	178.3
350	350	121.7	243.4

Source: Koluvek (1970)

There has often been little attempt to standardize selection and manufacture of the valve. Where this has been left to the contractor the valve quality is disappointing. Repair and replacement of valves will continue to be an important annual task so that valves should be made of durable materials which can be dismantled easily and for which spare parts are available or can be simply manufactured.

- The valve plate should be designed to fit into a recess in the top of the valve body to provide the seal, so avoiding the use of rubber washers which do not last and are difficult to refit. If surfaces are well machined the seal should be good.
- The most common maintenance problem is the rapid rate of wear of the cast iron centre in which the spindle turns. After several years of operation the thread becomes stripped and the spindle moves freely without turning, leading to continuous valve leakage.

  Two strategies which are being evaluated in an attempt to develop a more durable valve, which remains relatively low cost, include:

  i) use of an alternative non-ferrous metal centre piece; for example, gunmetal or bronze;

  ii) use of a removable insert which can be replaced when worn.
- Opening and closing of the valve using a key generally increases the time taken for valve closure and the likelihood of unauthorized rapid valve closure causing water hammer is much reduced. Slow valve closure is achieved by ensuring a small pitch screw thread on the spindle.
- A slip pin should be fitted in the tail of the spindle to prevent the valve top from being completely detached. This prevents damage to the valve lid by removal during normal operation, but allows easy removal when maintenance or replacement is required and is a better alternative to a lock nut which can become immovable due to rusting.

Construction recommendations

- Elaborate brick or masonry enclosures designed to prevent tampering with valves, as used on uPVC pipe systems built in Uttar Pradesh (Cunningham, 1986), tend to make repair expensive and difficult, and should be avoided. Although reducing the incidence of tampering, these enclosures have not been successful at eliminating damage to the valves. The extent of farmer involvement and co-operation seem more important influences on the frequency of damaged valves.

Table 5.5 **Recommended flow capacities for standard orchard valves**

Inside dia. of riser (mm)	Valve outlet dia. (mm)	Design capacity l/s Low head (0.3m)	High head (0.75m)
150	37.5	1.1	2.2
150	62.5	3	6
150	87.5	6	12
200	125	13	26
250	150	19	38
300	200	33	67

Source: Booher (1974)

- Where water is released directly into an earth channel the top of the valve should be set some 50 to 150mm below the ground surface to reduce erosion from the irrigation stream and minimize interference with farming operations. In these cases care should be taken to avoid high valve head losses as the turbulence may increase erosion around the valve (Koluvek, 1970).
- Where available head is low, the riser and alfalfa valve should be the same diameter as the pipeline to minimize the head loss.

Examples of valve leakage caused by wear and damage to outlet valves is illustrated in the publication 'Photographs of Buried Pipelines for Surface Irrigation.

Orchard valve

This is an outlet valve installed inside a riser pipe with a similar adjustable lid or cover to that in an alfalfa valve. Because the valve opening is smaller than the inside diameter of the riser its flow capacity is less. A typical orchard valve design and setting is shown in Figure 5.21 (Baudequin et al, 1990).

The valve may be installed at or just below ground level or may extend above ground and incorporate slide gates or notches. With their smaller flows these valves are often used in furrow irrigation of tree crops, from which the name orchard valves derives.

Recommendations regarding their design and construction are much the same as for alfalfa valves. Recommended sizes and maximum flow rates for orchard valves are detailed in Table 5.5.

Open pot outlet

The outlet consists of an orchard valve installed in the top of a riser with a section of larger diameter pipe mortared to the pipe and extending above it. The water is released through two or more outlet gates mounted on the sides of the pot (see Figure 5.21). This outlet distributes water to furrows and is mainly used in orchard irrigation. If the pressure is sufficiently low so that the pot will not overflow, the orchard valve can be dispensed with but water flow must then be controlled at the block level.

Recommended design capacities for open pot outlets are detailed in Table 5.6.

Capped risers or pot outlets

This valve comprises a riser extending above ground with a watertight cap and outlet gates in the side, positioned slightly above the ground surface (see Figure 5.20). An alternative

Table 5.6 Design capacities of open pot outlets with orchard valves

Dia. of opening (mm)	*Max. capacity (l/s)*	*Inside dia. of riser (mm)*	*Valve dia. (mm)*	*Flow cap (l/s)*
25	0.6	150	40	1.1
40	1.1	150	60	3.4
50	1.1	150	90	6.5
75	4.4	200	125	13.0
100	7.4	250	150	15.8
125	11.6	250	160	22.0
150	17.0	300	200	33.4

Source: Koluvek (1970)

to this capped riser is to install a pot with a diameter larger than the riser pipe in order to house more gates, and this is called a capped pot outlet.

The streams of water from the outlet gates can be erosive if pressures exceed 1m above ground level. The main advantage of capped outlets is they avoid debris clogging the outlet gates and the structure is simplified with the removal of the orchard valve. Like the open pot outlets these outlets commonly supply furrows for irrigation of orchards.

Maximum design capacities for slide gates on capped risers are detailed in Table 5.7.

End capped risers

This alternative riser comprises a 45° pipe bend of galvanized steel pipe, flange-jointed to a pipe tee and riser. The riser protrudes above ground by 0.2 to 0.3m and is closed either by a screwed end cap or a flat plate which is bolted to a flanged end. This type of outlet is common on small low-pressure pipe systems in Maharashtra, India, where the component parts are available from local irrigation equipment suppliers off the shelf. One of these outlets is illustrated in the publication 'Photographs of Buried Pipelines for Surface Irrigation.

Table 5.7 Pot sizes recommended for various sizes and numbers of slide gates

Dia. of opening (mm)	*Maximum number of gates*	*Size (mm)*
150	2	75
200	4	25
200	2	125
250	4	50
250	2	150
300	6	25
300	4	50
350	6	40
350	4	75
400	8	25
400	4	100

Source: Koluvek (1970)

Open stand with side orifices

These outlets comprise a pipe stand with simple orifices or pipes fitted with flanges in the side of the stand. Operation of the outlet relies on balancing the flow from the pipeline with the demand at the outlet. Typically the flow is controlled using an in-line gate valve.

Outlet distribution structures

A wide range of different designs exist for distribution structures associated with outlets for buried pipe distribution systems. These structures are considered a requirement only where the outlet flow is to be distributed via earth channels. Selection should be based on the functional requirements of the structure. The alternatives can be categorized according to their function and the materials and construction method used. These can be summarized as:

- earth-bunded outlets;
- simple concrete or masonry apron;
- concrete or masonry stilling basins with or without flow division;
- concrete or masonry stilling basins with flow measurement weirs as well as flow division.

All distribution structures must function to allow flow to be directed into one or more field channels as well as protecting the surrounding area from scouring if high velocity flows are involved. The structure must be of adequate capacity for the peak discharge which is envisaged, which may not be the same as the design flow. Where overtopping of the structure occurs, scour and undermining can quickly result and be expensive to remedy.

Although outlet discharge can be split from any outlet, the provision of accurate flow division will be invaluable wherever regular operation is to involve the sharing of the outlet discharge between two irrigators. Outlet distribution structures are many and various and some of those currently in widespread use are illustrated in the publication 'Photographs of Buried Pipelines for Surface Irrigation'.

Earth-bunded outlets

These represent the simplest and lowest cost alternative where farmers break and reform bunds as necessary to direct the water to particular field channels (Figure 5.22, type II). They cannot be used where outlet velocities are high and / or soils are not very cohesive and easily eroded. In positions where flow drops onto an adjacent field it may be necessary to provide a special drop structure.

Concrete or masonry stilling basins with flow division and flow measurement

The provision of a flow measurement facility is useful where the irrigator is involved in setting the outlet discharge within an agreed schedule of supply (for example, if two outlets are being used at the same time). A simple weir provides a visual check of the adequacy of the rate. Flow measurement at the outlet for the purpose of setting the correct valve opening or estimating water use is not likely to be practical and is only of direct interest to researchers and designers. Experience suggests that variable construction quality is also likely to result in weirs with inaccurate dimensions which at the very least will require individual calibration.

i) Alfalfa valve outlet

ii) Orchard valve outlet

iii) Orchard valve with open pot outlet

Figure 5.21. Alternative outlets 2

Source: Baudequin et al (1990)

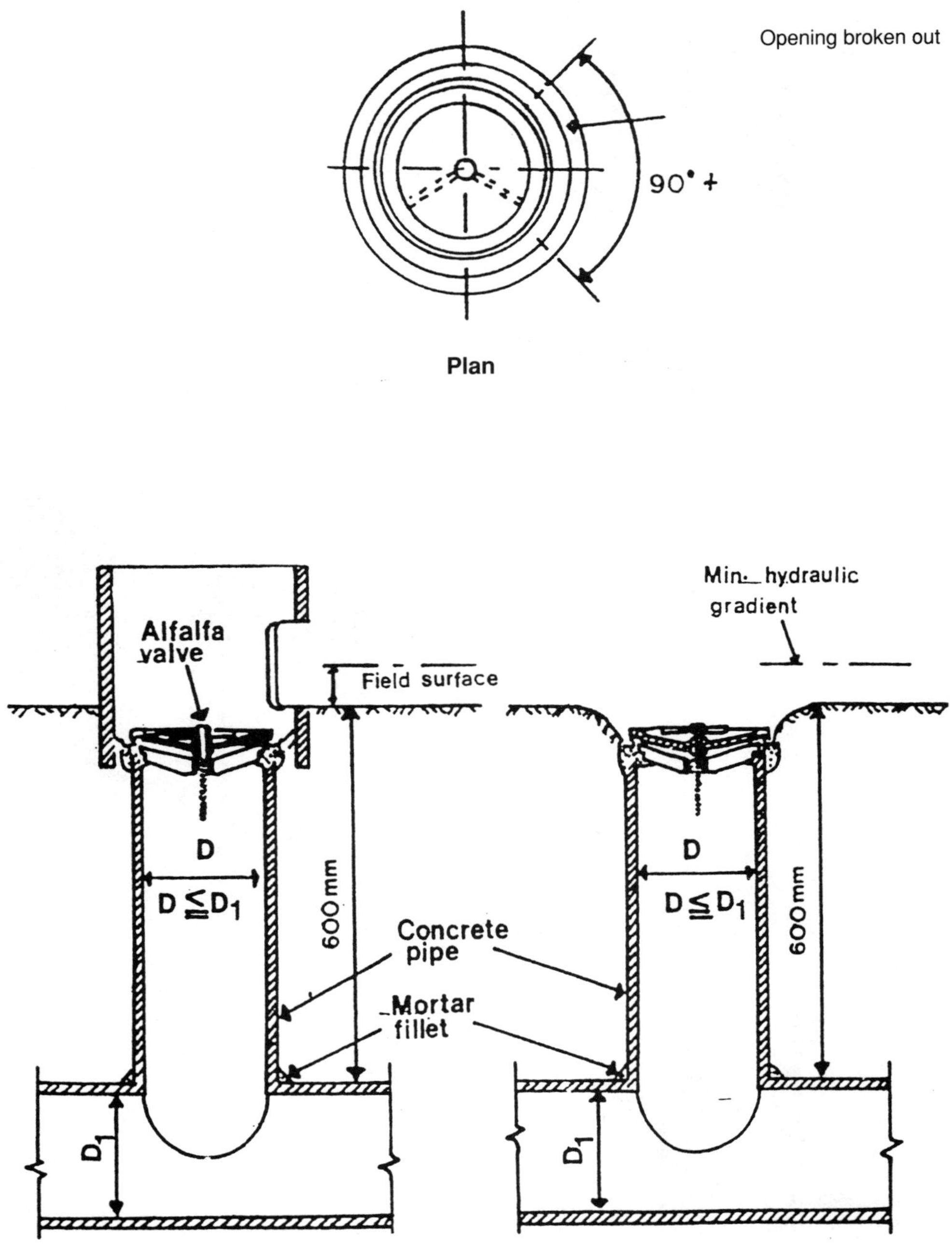

Cross-section
Type 1
Recommended when velocity in riser exceeds 1m per sec. and hydrants are not used

Cross section
Type II
Recommended when velocity in riser is less than 1m per sec. or when hydrants are used.

Figure 5.22. Cross-sections of alfalfa valves

Source: US SCS (1967)

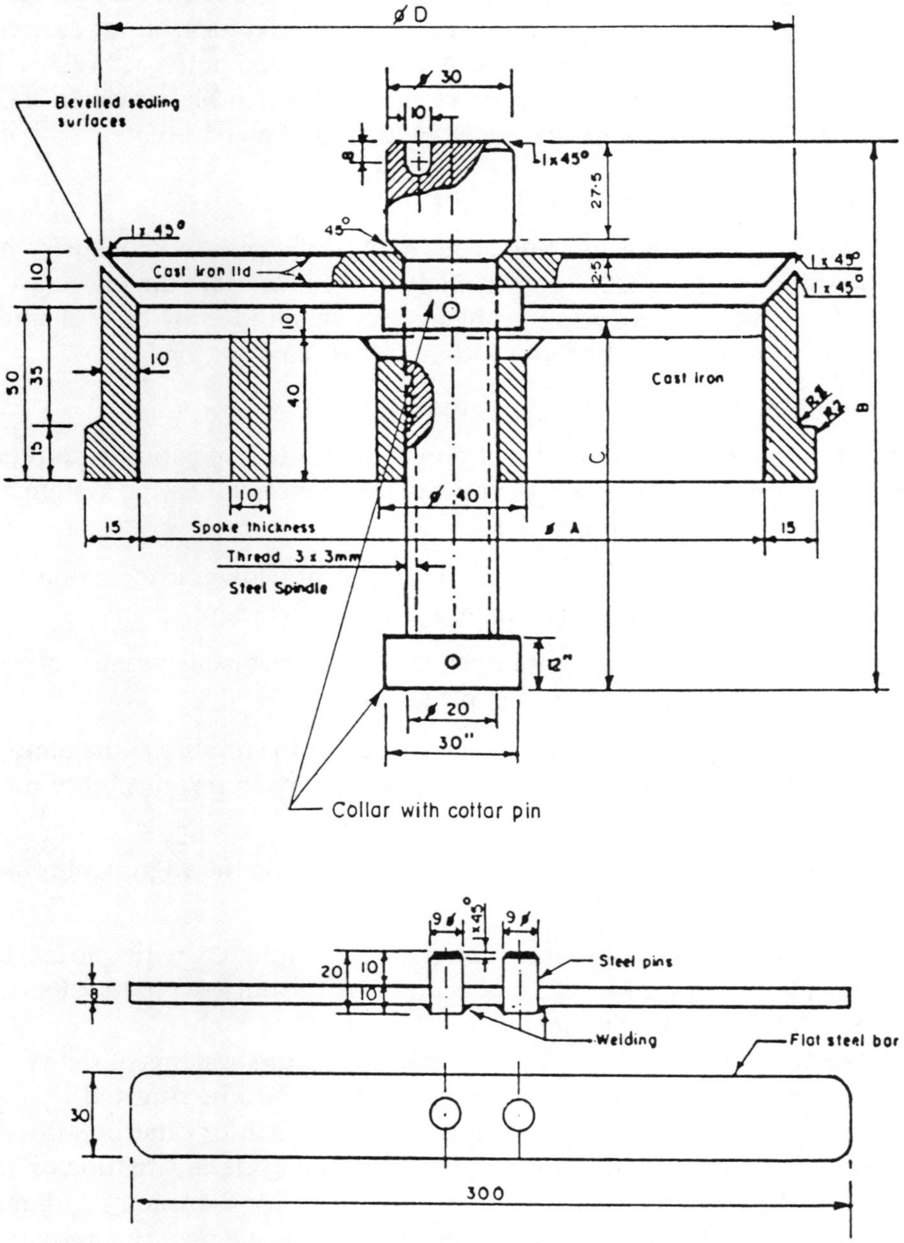

Valve size	A (mm)	B (mm)	C (mm)	D (mm)
150mm	150	150	105	170
200mm	200	150	105	220
225mm	225	160	115	245
250mm	250	160	115	270
275mm	275	170	125	295
300mm	300	170	125	320

Figure 5.23. Outlet or riser valve — alfalfa type Mark II

Source: MMI (1992)

The alfalfa valves must be set some 50 to 150mm below ground level or earth channel invert level to prevent excessive erosion. Farmers break and reform bunds as necessary to direct the water to particular field channels. An earth-bunded outlet as built by TADP in Bangladesh is detailed in Figure 5.19. In small pipe systems in Maharashtra State, India, irrigators use simple protruding risers closed by means of an end plate.

Simple concrete outlet box or masonry apron

Where outlet flow velocities are high and scour likely, a simple apron or surround can be adequate though side walls which direct the flow into one or more directions are likely to be essential. Many potential scouring problems can be avoided by careful siting of the outlet. This type of structure is the most widely constructed.

- Outlets should be set no higher than is absolutely necessary.
- Existing networks of field channels will provide a guide to positioning outlets. Small changes to outlet position will not affect the performance of the pipe system but could strongly influence the life of the outlet.
- The outlet should be set back from any field significantly lower in elevation and where that is not possible a drop should be included in the structure.
- Simple outlet boxes fitted with a simple weir can provide a visual measure of outlet flow for the farmer as well as protecting against scour.
- Where practical, individual outlets for each irrigator can simplify management, eliminate the need for large elaborate structures and apportion responsibility for maintenance.
- Outlet positioning should include obtaining agreement on the alignment of associated earth channel networks.

The most common design of structure comprises a simple box with one exit orifice or weir position. This type of structure has been used on buried pipe systems used in Thailand, India, Sri Lanka and Spain.

The outlet distribution structures built on the uPVC pipe systems of the World Bank Tube-wells Project in Uttar Pradesh, provide for the flow to be directed in any one of several directions. Flow division is not envisaged and earth or other readily available material is used to direct the flow. An alternative design under evaluation on the Indo-Dutch Tube-well Project, and earlier used on outlets in TADP systems in Bangladesh, comprises a concrete surround with two gaps for water movement in alternative directions. This will work well only where the outlet pressures and flow velocities are low. Examples of various alternative outlet distribution structures in use are illustrated in the publication 'Photographs of Buried Pipelines for Surface Irrigation'.

Concrete or masonry stilling basins with flow division

This type of distribution structure is widely constructed. Typically the alfalfa valve discharges in the centre of the structure or within a stilling well and water is directed to particular earth channels by a number of techniques including wooden slide gates, earth fill or concrete "pucca nuccas". The pucca nucca is a concept developed in Pakistan which incorporates a precast concrete plate with standard orifice sizes according to design discharge (Trout and Kemper, 1980). The orifice is closed by fitting a precast concrete lid.

A design for this type of structure is shown in Figure 5.24, as built by IDA DTW II in Bangladesh to distribute the flow between three field channels. Variations on this design where flow is distributed between two or four field channels have also been built.

Measurement of the entire outlet discharge may be by either a V-notch weir for smaller flows, or a broad-crested rectangular weir, which is less accurate but able to accommodate larger flows with a minimum of head loss.

An IDA DTW II project design of a distribution structure incorporating a weir, several of which have been built on pipe systems in Bangladesh, is shown in Figure 5.25.

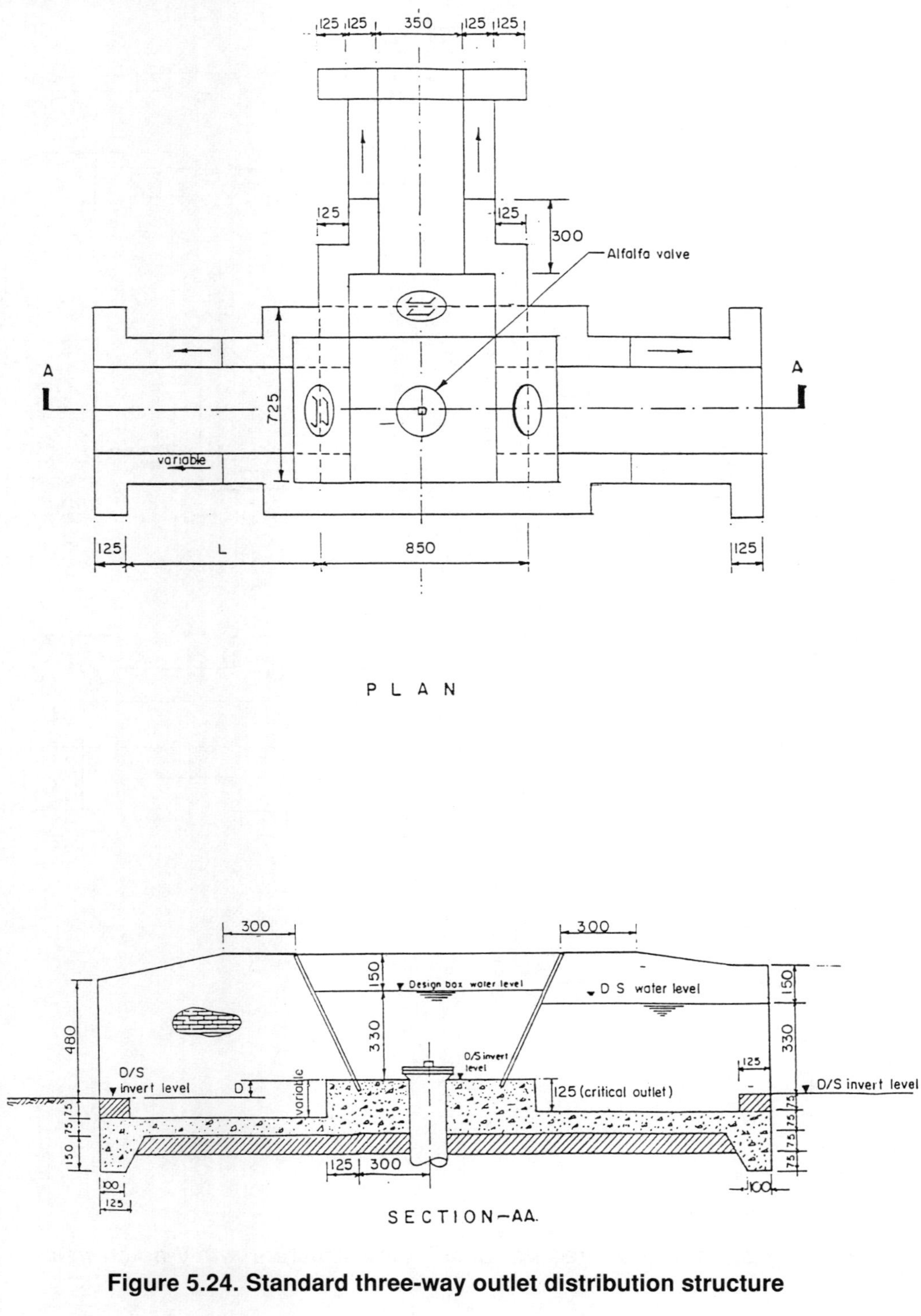

Figure 5.24. Standard three-way outlet distribution structure

Source: Sir M. MacDonald and Partners (1989a)

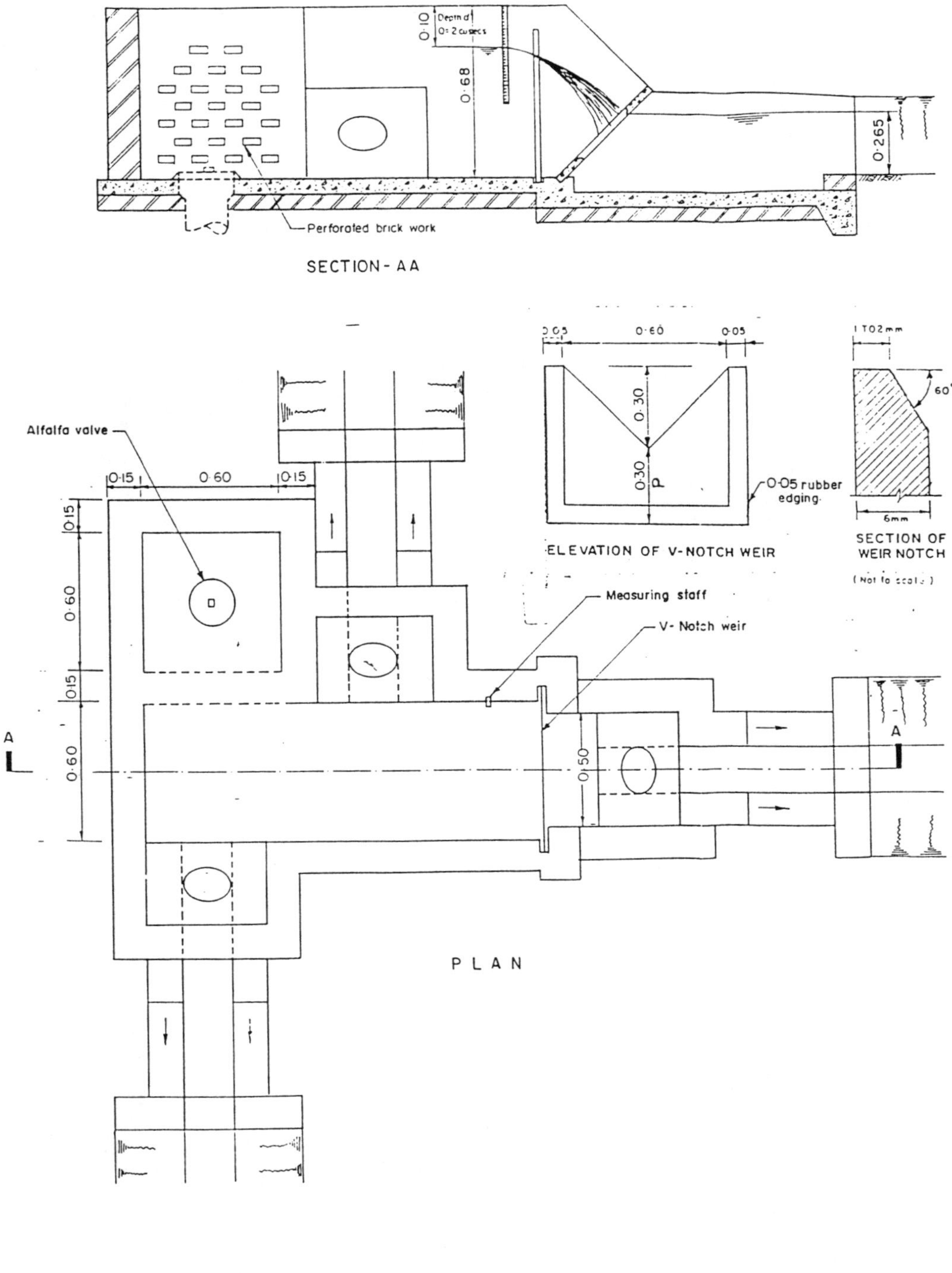

Figure 5.25. Standard three-way distribution structure with V-notch weir

Source: Sir M. MacDonald and Partners (1989a)

CHAPTER 6

SYSTEM CONSTRUCTION

Good construction practice is essential if pipelines are to be reliable and free from excessive maintenance and repair costs. Specific standards apply in different countries with respect to pipeline construction and this discussion does not seek to diminish their importance, but rather summarizes sound engineering practice together with relevant field experience. The following discussion draws heavily on experience and references received from IDA DTW II Project and Tangail Agricultural Development Project in Bangladesh and in particular from Georgi (1989), and Sir M. MacDonald and Partners Ltd (MMP 1989a).

In general, rigid pipe materials and plain-walled uPVC pipe will be suitable for low-pressure irrigation under a wide range of soil and loading conditions. References to methods of analysing the load limits for particular pipes are given. However, for thin-walled plastic pipe materials, careful design is necessary to ensure the pipe meets the installation and operating requirements. Apart from considerations of internal hydrostatic pressure, operating temperature and positive and negative pressure surges due to water hammer which have already been outlined, it will also be necessary to assess pipe-buckling strength under static and live loads and when subjected to external hydrostatic pressure.

Materials handling

Non-reinforced concrete pipe

In all cases pipe should be well cured before transportation and reasonable care taken to avoid unnecessary handling. Although pipes are generally manufactured in lengths of 1.8m, shorter 1m lengths allow pipes of greater than 350mm to be installed by hand. The heavier the pipe the more likely damage becomes through mishandling.

Spigot and socket pipe

Non-reinforced concrete pipe is particularly susceptible during rough handling and transportation, which can cause cracks or breaks particularly at the socket end of spigot socket pipe. The pipe frequently cracks at the neck because of the bending moment developed particularly with longer pipes. Pipe must be fully cured before being transported to site, and on arrival should be carefully inspected to identify any damaged pipe. This should be clearly marked and set aside. Socketed pipe should be stored upright with the bell-mouth or socket uppermost.

Plain-ended or tongue-and-groove concrete pipe

Because these pipes can sit horizontally on a truck deck or tray, breakage problems are reduced. Where care is taken pipes can be transported over considerable distances (up to 100km), with breakage of less than 1% (experience in Maharashtra, India). In general, though, breakage of up to 5% (reported in Bangladesh) should be expected with pipes of around 2m length and where only average care is taken.

Measures which can help to reduce breakage during transport include:

- covering the truck deck with 100mm depth of sand and placing packing such as used sacking (or jute bags) between adjacent pipes;
- use of shorter lengths of pipe which are less prone to cracking because of the lower tension forces developed during handling, although the number of joints are increased.

When stored at site plain-ended and tongue-and-groove pipe should be stacked no more than four pipes high and wedges must be placed to prevent the lower pipes from separating.

uPVC

Damage to the pipe wall and in particular the spigot and socket joint can affect the pressure rating of the pipe and possibly its service life. Exposure to sunlight for any period, except for specially formulated pipe, requires that the pipe be derated according to manufacturers' recommendations. Careful handling, transport and storage are essential to ensuring pipe quality does not deteriorate. Good storage practice is illustrated in Figure 6.1.

Weatherability

High temperatures

At high temperatures the impact strength of uPVC pipe increases but with a corresponding drop in hoop strength, which is a measure of the strength of the pipe to resist internal and external pressures. Therefore as the sustained ambient or fluid temperature increases above 20°C the pressure rating of the pipe should be reduced.

Manufacturers supply pressure-derating tables which should be used. In the absence of these, a conservative rule of thumb is to reduce the pressure rating by 2% for every degree Celsius above 20°C, up to a maximum acceptable temperature of 60°C.

Some recent work suggests that this derating may be too conservative for temperatures up to about 33°C. These results may have significance for many developing countries with hotter climates (WRC, 1989).

Sunlight

WRC (1989) describes two ways in which direct exposure to sunlight affects uPVC pipe.

Firstly there is a "solar gain" where the pipe temperature can rise above the surrounding air temperature. This will tend to cause a greater risk of pipe deformation, such as ovalization due to stacking or bending deformation when not fully or evenly supported.

Secondly exposure to strong ultraviolet light can cause a thin film of degradation on the exposed surfaces of uPVC pipe, to a depth of 0.05mm. This is significant for thin-walled uPVC pipe materials. In an advanced stage this is seen as a discoloration of the pipe and, at a micro-level, surface effects are thought to be possible sites for the initiation of cracking, and can contribute to a reduction in impact strength of the pipe.

WRC (1989) describes some of the practical results of this as follows:

- UV exposure results in colour change, a slight increase in tensile strength and reduction in impact strength, hence the comment that pipe becomes more brittle.
- UV degradation only happens on the uPVC pipe surface directly exposed to radiation, penetrates to only a shallow depth, and stops as soon exposure ceases.
- Exposure of the inside of the pipe at the ends is of more concern than that of the outside of the pipe.

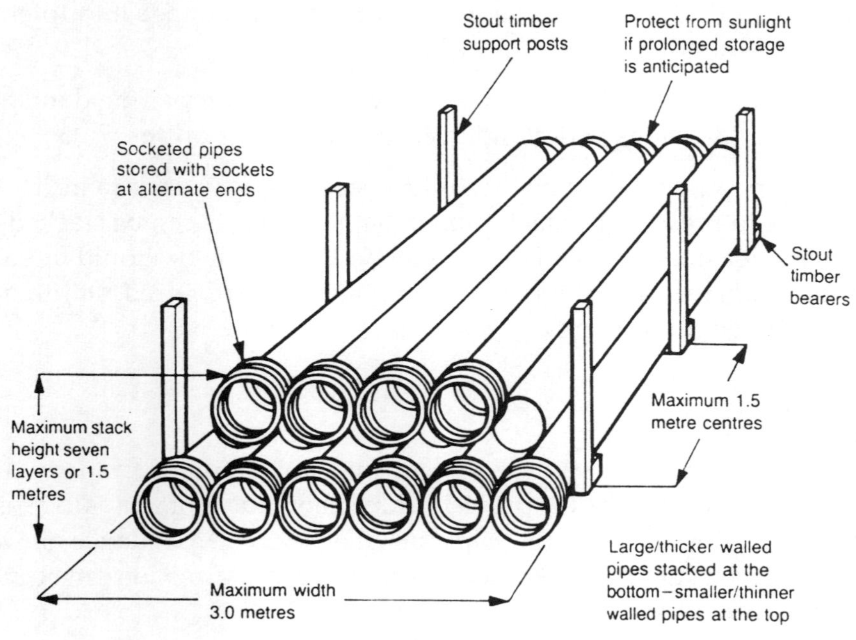

Figure 6.1. Pipe storage recommendations

Source: Sir M. MacDonald and Partners (1989e)

WRC (1989) concludes that thin-walled uPVC pipe materials, normally containing a pigment block such as titanium oxide, will be protected during long periods of UV exposure, non-pigmented pipe should be kept out of strong direct sunlight, although tree shading will usually be adequate. The rate of deterioration of the pipe surface will depend on the light intensity.

Good procedure states that:

- Pipes should not be dropped, dragged or struck with a hard object.
- Pipes should be stacked by hand and special storage arrangements made at the site.
- Pipes should be stacked on hard surfaces so they are evenly supported along their entire length.
- When stacked, respective layers of pipes should have the spigots at alternate ends and the stack should be covered. A tarpaulin will protect the pipe from damage from sunlight which can cause both surface degradation and changes in shape, such as longitudinal bending. For short periods, storage in heavily shaded locations can be adequate.
- Pipes should be delivered to site a minimum of time in advance of being required.

Asbestos cement pipe

Much of what is correct procedure should be common sense, but it is worth at least mentioning that:

- All pipe and fittings must be checked on arrival for damage and to ensure that the delivery corresponds to what was ordered.
- Careful unloading is essential to avoid extra work later repairing damaged materials. Use rope or nylon slings and do not roll or throw joints off the truck.
- Like other pipes they should be carefully nested, but no higher than 2m, on a flat base with wedges to prevent the stack from collapsing. Joints can be stacked three or four high for smaller sizes or rested on each other. Rubber rings should be stored in a cool shady place, together with the jointing compound until required. Oil, grease and strong sunlight damage rubber rings.

Setting out

Ideally pipelines should be straight and of uniform gradient from reach to reach or between stands. Bends require the provision of anchor blocks or straps to resist axial forces on the pipe and this adds to the pipe installation cost. Guidance on anchor block construction is given later in the chapter. Large changes in direction can be made at stands or similar structures in some cases.

Non-reinforced concrete pipe

Non-reinforced concrete pipe does not easily accommodate changes in direction, and joints are more difficult to make watertight where gaps between pipes are uneven (for example where socketed pipe is used on a bend) or large (greater than 10mm). Where plain-ended pipe is used bends are made by chipping one side to create the required angle. These joints are easily made but can prove troublesome, so as few as possible should be used.

Manufactured bends are available in some situations but are often avoided as they are expensive and need to be ordered in advance in the required quantities and dimensions. Waiting for proprietary fittings or bends can mean expensive delays.

An alternative bend concept, used in the USA and worthy of consideration where tongue-and-groove pipe in particular is being manufactured, is found under the trade name of Rotacurve. It comprises a pipe pair manufactured as part of the same mould which has an offset tongue-and-groove or spigot socket joint. This allows the pipes to be rotated relative to one another to assume different angles of bend. As a pipe pair can also be laid straight excess pipes are not wasted. This technique allows broad bends to be constructed, so minimizing the thrust forces developed (Rotacurve).

uPVC pipe

In general, uPVC pipe should be laid in as straight a line as possible and although uPVC pipe can accommodate considerable bending this should be avoided except where it is necessary to avoid obstacles. Manufactured fittings are commonly used for sharp changes in direction and require anchor blocks in all cases.

In general, cold bending of uPVC pipe should not be carried out for pipes larger than 160mm in size because of the additional stresses on the pipe.

For these smaller pipes the cold bending radius (R) of a uPVC pipe length should not be less than 200 times the external diameter.

R_{min} × External diameter

For a pipe of 160mm this represents a radius of 30m or a deviation of 0.5m on a six-metre pipe length. The pipe should be securely fixed using anchor blocks and straps at the spigot and socket ends as well as in several intermediate positions.

Bending should not be done in cold temperatures, below about 5°C.

Excavation and design for pipeline loading

The loads exerted on buried pipe are related to the laying condition, the type of bedding, the degree of compaction achieved and the nature of the bedding material.

Pipeline design for rigid pipes, concrete and asbestos cement

The theory of calculating loading on buried rigid pipes is well explained in a number of texts and usefully summarized by Young et al (1983).

In practice manufacturers provide summary tables for their pipe material which give recommended minimum and maximum cover depths for particular trench conditions, backfill materials and pipe diameters.

The following section describes some of the commonly discussed concepts. Rigid pipes are subjected to two types of forces during loading:

- crushing or ring bending;
- beam action or longitudinal bending.

Crushing

When considering a pipeline in cross-section, the forces which produce bending moments in the pipe wall come from three sources:

- the weight of the overlying fill;
- the soil pressures transmitted down to the pipe from surface loads;
- the supporting reaction of the bed of the pipe.

For safe design the maximum total forces within the pipe wall must be within the limits that the pipe can support over its expected life. It is now generally accepted that the strength of rigid circular pipes should be determined from two or three edge bearing tests on sample pipes rather than using bending theory.

Longitudinal bending

Failure in bending is usually confined to rigid pipes of 300mm or smaller external diameter. Poorly supported, shallow buried rigid pipes below roads can develop large bending moments.

Conditions of installation

There are four main situations in which pipes are installed:

- in a narrow trench which is subsequently backfilled;
- in a wide trench or on the surface of the ground over which an embankment is subsequently built and termed the positive projection condition;
- in a trench in the ground over which an embankment is subsequently built and called the negative projection condition, and
- in a tunnel or heading by pipe-jacking.

In general, for the situations of interest in buried pipe systems used for irrigation, only the narrow trench and the positive projection conditions will be found. For the other two conditions, reference should be made to Young and O'Reilly (1986) for the required calculations. Manufacturers usually provide recommendations which should be followed for the depths of cover for their pipe material.

Laying condition
Apart from the situation where the pipe is exposed above ground level and carries no load, the two conditions of interest are trench and culvert. In most cases pipe will be laid in a narrow trench and the weight of fill which impinges on the pipe is less than the total weight of fill material. The trench width should be as narrow as possible, as where it exceeds four times the diameter of the pipe, the culvert or embankment condition applies, so increasing the load on the pipe.

The culvert condition is where the pipe is laid in a shallow excavation (under embankment) and the pipe projects above the adjacent undisturbed material. Under these conditions the vertical load transmitted to the pipe is usually greater than the weight of the fill material above the pipe because of the transfer of load from adjacent fill through cohesive drag.

In some parts of pipe systems the culvert condition may exist and need to be considered.

Type of bedding
The type of bedding typical in low-pressure buried pipe systems is ordinary projection bedding where the pipe is carefully supported in the bottom of the trench for at least 50% of its diameter, and granular bedding material is carefully compacted around it.

Degree of compaction and bedding material
Very fine-grained soils, those with high plasticity, and organic soils are all unsuitable for bedding and haunching of pipelines, and as initial backfill material.

Pipe load design for non-rigid pipes: uPVC pipe

uPVC pipes need to be treated differently to rigid materials, because they carry a lower proportion of the applied load. uPVC mobilizes passive forces within the bedding material supporting the sides of the pipe.

The standard approach to design has two aspects:

- ensuring that the deformation of the pipe under load remains within acceptable limits; and
- ensuring the pipe does not fail due to buckling.

Pipe deformation

To ensure satisfactory long-term performance of low-pressure uPVC pipelines the long-term deformation should be restricted to within 6% of the vertical diameter of the pipe.

The pipe deforms by an increasing amount over its life but most of this occurs within the first two years after pipe installation.

The deformation of uPVC pipes can be calculated using the Spangler "Iowa" formula:

$$\text{Deformation} = \frac{\text{Vertical load on pipe}}{\text{Pipe stiffness} + \text{Soil stiffness}}$$

Procedures for calculating the deformation are described by the Water Research Centre (WRC, 1986).

The pipe, when empty at installation, will be subject to some deformation. It is important to carry out a design check to ensure the vertical deformation of the pipe is well within the maximum allowable deflection of 6% and that the pipe is safe from buckling failure.

Bedding and compaction

Very fine-grained soils, those with high plasticity, and organic soils are all unsuitable for bedding and haunching of pipelines and as initial backfill material. Particularly in the case of thin-walled uPVC pipe materials, granular material with little or no fines requires less compaction to achieve an adequate density. Where granular or non-cohesive soils are used reference should be made to manufacturers' recommendations regarding the load capacity of their pipe material, for example Jain Ribloc (1991).

Excavation

To keep the quantity of excavation to a minimum the trench should be excavated with straight sides. Where the trench is particularly deep or the soil is unstable a slight batter on the trench walls may be necessary to avoid material falling into the trench before or during construction. In some coarse-textured soils, some type of temporary formwork may be necessary.

Excavation should be avoided in heavy rain as this may cause the trench walls to collapse inwards. Water in the trench should be removed if possible and the trench allowed to dry out before pipe-laying. Trench excavation should be started a minimum of time in advance of the pipe-laying, so as to minimize the disruption caused to normal agricultural operations. The excavated material should all be placed on one side of the trench if possible to enable easy access to the pipe from the opposite side.

Ideally the trench should be free from clods, stones, obstructions and very soft areas. If the trench is excavated in rock, boulders or other hard material or through soft areas, it should be over-excavated by at least 0.3m and filled to grade with whatever suitable material is available locally. As this can add considerably to the cost all attempts should be made to find an alternative alignment or solution.

Particular problems can occur in soils which shrink and expand with changes in moisture content or temperature. Increasing the depth of excavation is usually sufficient to avoid the problem. However, in some situations it will be necessary to use a more flexible pipe material and an alternative backfill material to that in situ. The soil conditions should be carefully considered in the early stages of design, and the most important issues should be covered during pipe material selection as illustrated in Flow chart 2.8.

The trench should be of sufficient width to give adequate working space for pipe-laying and jointing. The minimum width will be about 0.5m for pipes under 300mm in diameter and for larger pipe sizes the trench will need to be at least 0.4 to 0.6m wider than the pipe. For larger-diameter pipe the bottom of the trench should be shaped to take the bottom half of the pipe.

Pipe cover

Unnecessarily deep excavation only increases the cost of installation and should be avoided. The required cover will depend on the agricultural practices and the pipe material. The pipes should be deep enough to avoid vehicle traffic, farming operations, freezing temperatures and soil cracking.

Typically, for agricultural land, a minimum depth of between 0.4 and 0.6m above the pipe crown is recommended for all pipe materials. Larger diameter pipe, and pipe installed in heavier (clay) soils, will require the greater depth of cover. Specific cover limits relating to the soil texture and type of traffic load are available from pipe manufacturers for uPVC pipe materials. In particular, thin-walled uPVC pipe should not be installed with deep cover.

For concrete pipe where high depths of fill are involved a safety factor of at least 1.25 should be applied to the three edge bearing test when calculating allowable external loads of fill material over the pipe (ASAE S261.7, 1989). Where pipes cross under roads or embankments, cover limits should be calculated assuming the culvert condition. For concrete pipe, greater depths of cover (0.7 to 0.9m) should be provided. This is frequently not a problem as roads are often on raised embankments above field level. Adequate compaction of the soil around the installed pipe during backfilling is very important to the load-carrying performance of a pipe, particularly in the case of uPVC pipe.

Thin-walled uPVC pipe

Specific recommendations will depend on the profile and stiffness of the particular pipe material, and manufacturers should be consulted for advice on their own pipe material.

As an indication, data from Jain Industries for their Ribloc pipe material is summarized here. Recommendations are provided on the load limits for the pipe installed in a trench at various embedment depths.

Although in situ soil or clay is not recommended as a backfill material around the pipe, it can be used given selection of a profile with sufficient stiffness. Profile selection for different surround materials is outlined in Jain Ribloc (1991). Given this selection, the following recommendations are made.

Minimum cover depths

Unless special backfill material is used requiring minimum compaction (for example, crushed gravel), a minimum cover of 0.75m is required for light traffic loads.

For a standard wheel load as defined by NAASRA A14 (Young and O'Reilly 1983), a minimum cover of 1.5m is required.

Maximum cover depths

For the range of pipe size from 225 to 750mm, recommended maximum cover depths range from 6m for the smallest diameter pipe to 3m for the largest (see Jain Ribloc, 1991). Where depths of cover exceed these figures, manufacturers should be consulted for the particular situation and design.

Pipe laying

Concrete and asbestos cement pipe should be lowered carefully into the trench without damage and placed in the centre of the trench. Ropes can be used to assist with this and keeping the socket end up will help avoid it being damaged. The pipe should as far as possible be in contact with a firm foundation along its entire length. As already mentioned the pipe should be seated in contact with the trench bottom for at least 50% of its diameter. For smaller diameter pipe (< 300mm) this will occur where the bed has been loosened or a cut provided. For joints involving a bell, collar or socket, space should be excavated below trench bed level so that the alignment of the pipe is not disrupted.

No pipe that is cracked or damaged should be used and any pipe showing serious settlement or misalignment should be taken up and re-laid.

uPVC pipe once joined at the side of the trench can be lowered into the trench while avoiding excessive bending.

During laying, dirt and foreign material should be kept out of the pipe. To assist with this a "pig" which consists of a stuffed bag tied with a rope is dragged through the pipe behind the pipe-laying team. The pig prevents large pieces of debris or tools from being left behind in the pipe and later causing blockages.

Air circulation through a pipeline should be minimized when it is being laid and when it is not in use. This helps to prevent the development of severe moisture gradients across the pipe wall. Blocking of the pipe openings, as well as minimizing air circulation, will also prevent vermin from entering the pipe after the day's work. Nests and drowned animals could initiate pipe blockage.

uPVC pipe

Pipe-laying should be avoided in extremely hot, cold or wet weather. This may mean only installing pipe during a particular part of the day. After pipe is laid and jointed, covering the pipe as soon as possible will allow the pipe to reach a uniform temperature before backfilling proper begins. This is particularly important with uPVC pipe, given its high coefficient of expansion. Although weather conditions cannot be predicted with any certainty, a short break may be enough until weather conditions improve.

Flexible uPVC pipe may develop a degree of distortion (oval shape) during storage or transportation after manufacture. So long as this is only in the order of 1 to 2 % of the pipe diameter it can be countered by laying the pipe with its maximum dimension in the vertical, so offsetting a certain amount of the deflection developing after installation (Jain Ribloc, 1991).

Non-reinforced concrete pipe

The hair-line cracks which sometimes occur in concrete pipe after installation are caused either by a drop in soil or water temperature or by drying out of the pipe resulting in tension forces being set up in the pipe. The most common cause is low-temperature water from surface supplies in winter and early spring. Well water is usually not a problem, being more even in temperature through the year. These cracks can be partially prevented by pre-stressing the pipe longitudinally. This tends to happen automatically when the pipe, on being laid dry, expands on becoming wet. Taking advantage of this pre-stressing and avoiding the development of excessive moisture gradients across the pipe will also reduce the frequency of other types of pipe and joint failure (Koluvek, 1970).

Jointing

Concrete pipe

Jointing systems in common use for low-pressure concrete pipe include:

- plain ended;
- spigot and socket;
- tongue and groove;
- rebated.

For some of these joint types there are also alternative methods of achieving the watertight seal. These methods comprise two main types:

- rigid cement mortar joints;
- flexible rubber ring or gasket joints.

Rigid joints use cement mortar to achieve the water seal, but this allows for little movement and when settlement occurs uneven loading can result in disruption of the joint and subsequent leakage. Various types of rigid joint are shown in Figures 6.2 and 6.3. Care needs to be taken during installation, curing and backfilling to avoid damaging rigid joints.

Completely watertight pipe systems using rigid, mortar-jointed non-reinforced concrete pipe are not possible, though particular pipe joint alternatives seem to perform consistently better than others.

Flexible joints use a rubber ring or gasket to achieve the water seal and allow a small amount of movement, depending on the design of the joint. These joints tend to be more expensive because of special requirements for the manufacture of the pipe. Examples are shown in Figure 6.4.

The availability and ease of manufacture of different types of concrete pipe in different parts of the world will affect the jointing options available.

Plain-ended pipe

Investigations into concrete pipe systems in Bangladesh using plain-ended pipe showed that most leakage problems occurred from joints between the header tank and the first outlet (i.e. where pressures are highest), or around structures, and particularly road crossings (Rashid et al, 1991). Special attention and supervision should be paid to these situations.

The two types of rigid joint used for plain-ended pipe are:

- the collar joint (Figure 6.2b);
- the sacking and mortar band (Figure 6.2c).

Collar joint

Use of a collar adds at least 10% to the cost of the total pipe installation, although it can provide a stronger joint to help withstand unbalanced soil loads.

If the collar gap is too small because of uneven manufacturing, poor specification or changes in pipe alignment it can be very difficult to place mortar to achieve a watertight seal. The joints are slow to make (typically 20 per manday) and leakage problems often occur where the standard of construction and supervision is poor. If an alternative is available the collar joint should not be used.

The collar is typically 150 to 200mm wide, often lightly reinforced, and has an inside diameter slightly larger than the outside of the pipe, allowing a gap of 15 to 20mm. Two variations on the joint are in use:

- The pipes are butted together and a stiff mixture of 1:2 or 1:3 cement:sand mortar is prodded into the space.
- A bitumen-soaked rope is placed in the gap between the two pipes and a bitumen-soaked sacking (or jute) band in the gap between the collar and pipes. Mortar is used to fill the final collar gap. This was the practice on pipe systems built by the Barind Integrated Agricultural Development Project (BIADP) in Bangladesh.

Jute and mortar band

Several variations of this joint are in use but the best results have been obtained with the following procedure:

- The two pipes to be jointed are wet and butted together.
- A small hollow is scooped out below the pipe to allow access to the underside of the pipe join.
- Cement slurry with a little sand is applied to a narrow (100 to 150mm wide) strip of thin sacking (jute) which is wrapped at least once around the join.

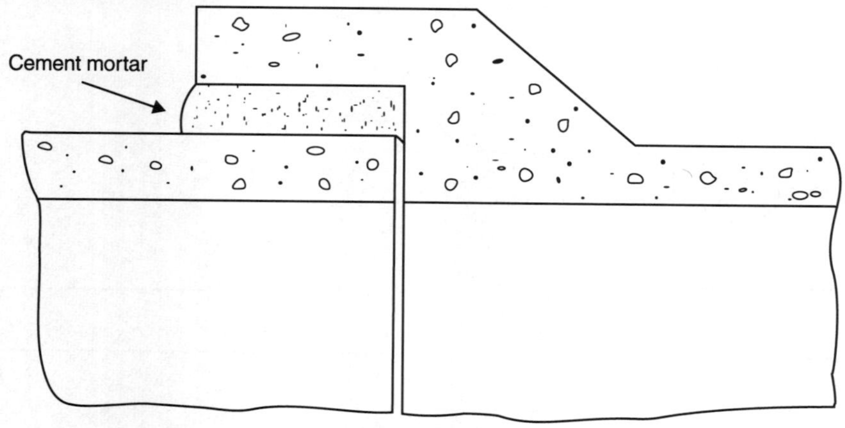

(a) A spigot and socket joint

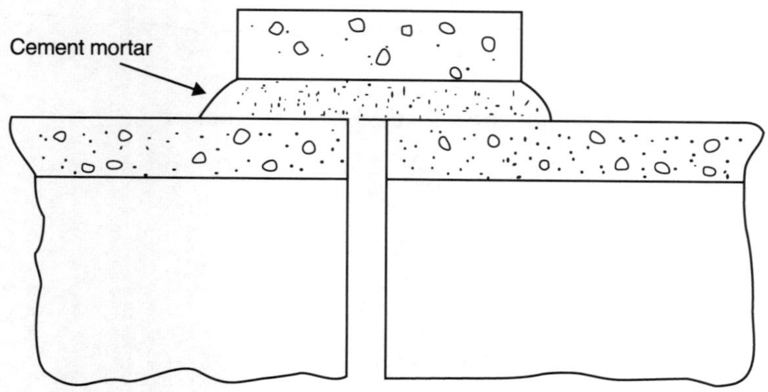

(b) A collar joint

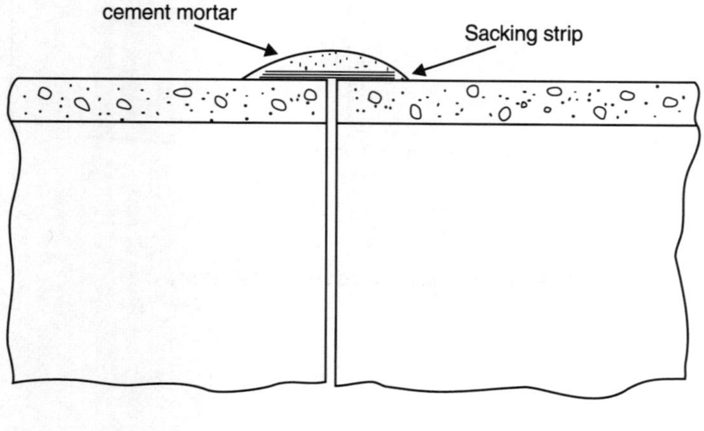

(c) A mortar band joint (plain-ended pipe)

Figure 6.2. Rigid joints for non-reinforced concrete pipes

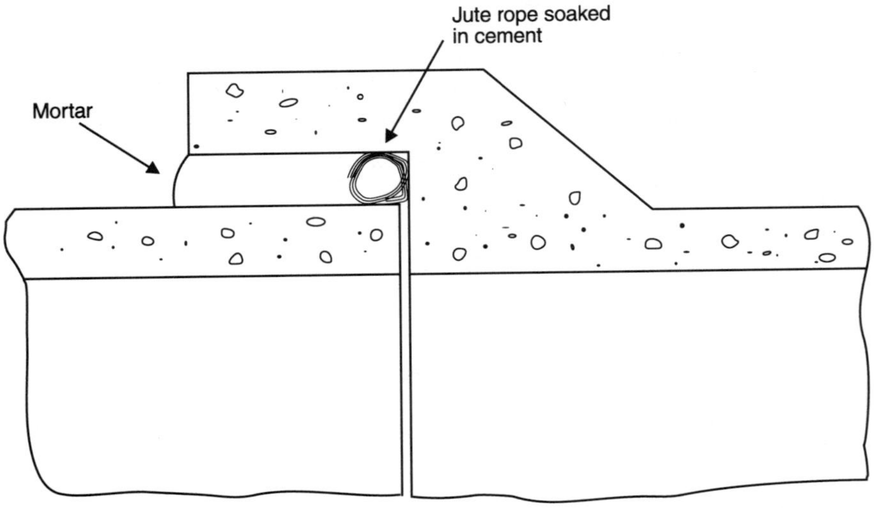

(a) A spigot and socket joint — alternative

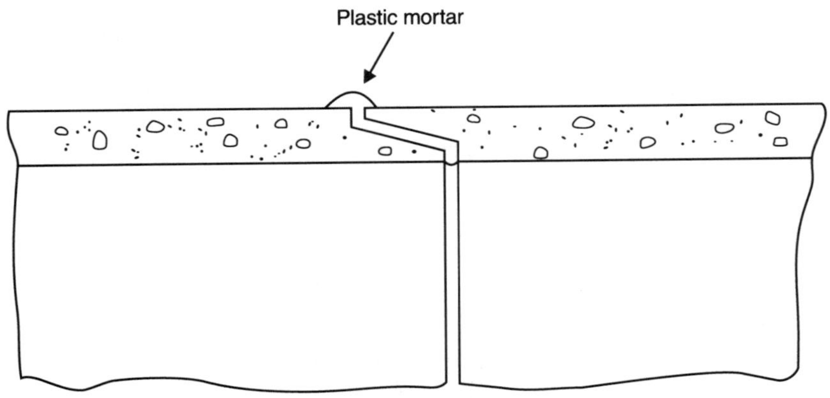

(b) A tongue and groove mortar joint

Figure 6.3. Rigid joints for non-reinforced concrete pipes (continued)

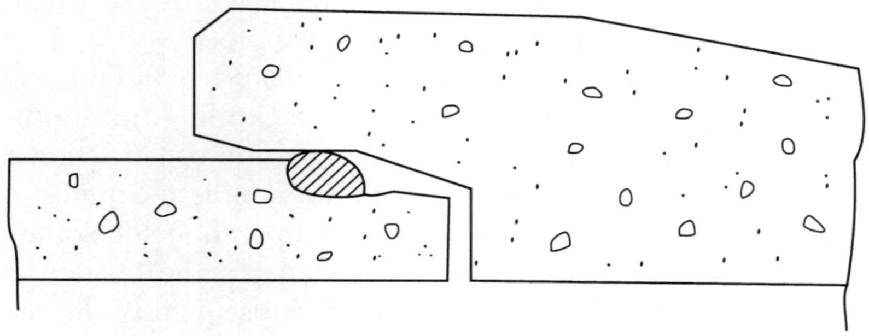

(a) Roll-on joint

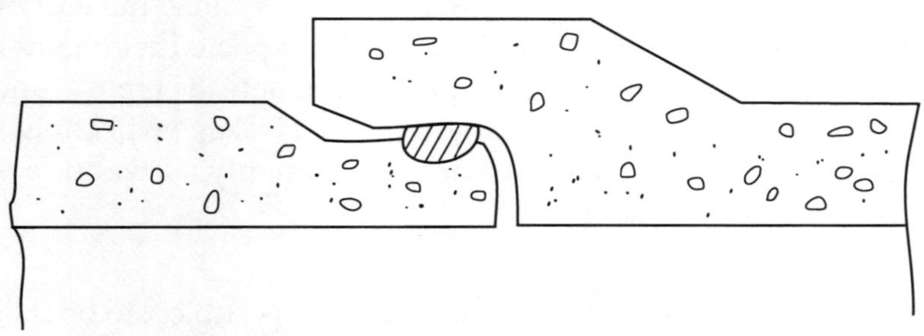

(b) Confined O-ring

Figure 6.4. Flexible joints for non-reinforced concrete pipe

Source: Indian standard 783 (1985)

❍ Once this cement-soaked sacking has dried (10 to 15 minutes later), a 100mm-wide layer (or band) of mortar comprising 1:3 cement:sand is applied (see Figure 6.2).

In the case of non-reinforced concrete pipe where plain-ended pipe is jointed with a mortar band over a sacking (or jute) bandage, a small hollow needs to be excavated under the join to allow mortar to be placed on the underside of the pipe.

The joint, though relatively cheap and quick to install (100 per manday), requires the pipe to be carefully bedded. Care needs to be taken during backfilling and curing to avoid any disturbance of the joints. Failure can occur due to poor joint construction as well as from pipe movement during settlement; however repairs are simple and quick to perform.

On TADP schemes in Bangladesh which have been operating for some time the performance of this joint has been mixed. TADP staff report that for the first few years leaks occur at the rate of 5 to 7% of the joints on the scheme. Subsequently the rate of leak occurrence depends very much on the quality of the original construction and jointing. This is supported by work reported by Rashid et al (1992).

Spigot and socket

Because of the special spigot socket joint, pipelines using this material are more expensive than where plain-ended pipe is available. The socketed joint has the advantage that it can allow gradual changes in pipe alignment to be taken up in the joint but not exceeding 3 to 5°. The pipe is difficult to manufacture using ordinary pipe-spinning technology and is usually made using automated vertical mould processes. Both rigid cement mortar and flexible rubber gasket joints are used with this pipe.

Cement mortar joint

Examples of this joint are shown in Figures 6.2 and 6.3. Because the end of one pipe is confined in the head of the next, fewer problems can be expected with joint leakage after installation and testing. However, there can be problems where the pipes are not in line with one another when filling the lower section of the joint, and leakage will often occur here.

A mortar with a ratio of 1:3 or 1:2 cement:sand is commonly used and sacking (or jute) fibre may be added to reinforce the mortar.

Flexible rubber gasket joint

These joints would be recommended where the pipeline is subject to occasional stresses requiring a limited flexibility. This could include seasonal changes in the temperature of the water or limited cyclic soil movement. Recommendations on particular gasket profiles to be used should be sought from pipe manufacturers. Examples of some of the types of rubber gasket joint in use are shown in Figure 6.4.

Pipe orientation while laying

Strictly the pipeline would be expected to perform more reliably in terms of being watertight if the socket is placed upstream against the prevailing flow direction. In practice it has been found easier to install in the trench (with fewer damaged pipes resulting) when the spigot end is inserted into the socket of the previously jointed pipe. Where pipe is laid beginning at the tube-well and working downstream the socket will be oriented in the opposite direction. There appears to be no strong evidence that one or other practice results in improved performance of the pipeline.

Tongue and groove

This joint alternative appears to be the preferable option for non-reinforced concrete pipe. It has been proven through use in the USA over several decades and is being successfully used now to a limited extent in India. An adequate standard of pipe manufacture is essential if this jointing technique is to be used successfully but tongue and groove pipe of adequate quality can be made on both mechanized vertical mould equipment and labour-intensive spinning equipment. Quick and reliable installation is possible and when achieved, leakage rates can be very low (Merriam, 1985). The frequency of leak occurrence on newly constructed pipelines in Maharashtra, India was barely 1% (as reported by site staff) compared with the 5% found in mortar-jointed plain-ended pipe in Bangladesh (MMI, 1990b). For larger diameter pipes, shorter sections can be made, to enable easier handling.

Reference has only been found to the use of this pipe with rigid mortar joints although rubber gasket joints with specially manufactured pipe are a possibility. Dimensions for the tongue and groove are detailed in Figure 6.5.

A variation on the standard pipe is manufactured to allow bends to be negotiated without necessitating special fittings. This comprises pipe made in pairs with alternate tongue and groove or bevelled bell and socket joints and is known as Rotacurve.

Rigid mortar joint

The recommended procedure is to place a plastic cement mortar into the groove of one pipe; the tongue of the other is pushed home, forcing out excess mortar. A soft brush on a long handle is used to smooth off the inside of the joint. The plasticity of the mortar is very important and comes from a proportion of fine material in the mortar. Workability of the mortar is more important than strength. The mortar can be considered plastic when it adheres to the underside of the pipe. The joint is often completed with a mortar band, usually without sacking, around the entire pipe circumference although this is considered cosmetic rather than functional (see Figure 6.4b).

It is preferable that the groove end of the pipe be pointing away from the water source.

uPVC pipe

uPVC is usually supplied in 5 to 6m lengths with spigot socket ends. The pipes are commonly jointed using a solvent cement compound although rubber ring seals can also be used.

It is important to remember that uPVC pipe is a very general term for a wide range of different plastics materials. For a particular pipe from a particular manufacturer it will be important to use the specified solvent-cement compound for jointing. Using old solvent stock or a product from another supplier may result in at best weaker joints and at worst joint failure. Test the jointing compound on a section of pipe before proceeding with the installation to evaluate its performance.

Solvent cement jointing

Although solvent cement joints are no longer used for jointing pipe larger than 150mm in higher pressure applications in the UK, most low-pressure pipe used for irrigation is still joined with this technique. Variations outlined below will occur with the type and specification of the solvent cement provided or recommended by different pipe manufacturers.

General principles

- The surfaces to be joined are pre-cleaned by wiping with a cleaning solvent, such as carbon tetrachloride. In practice wiping the surfaces with a clean dry cloth may be adequate preparation.

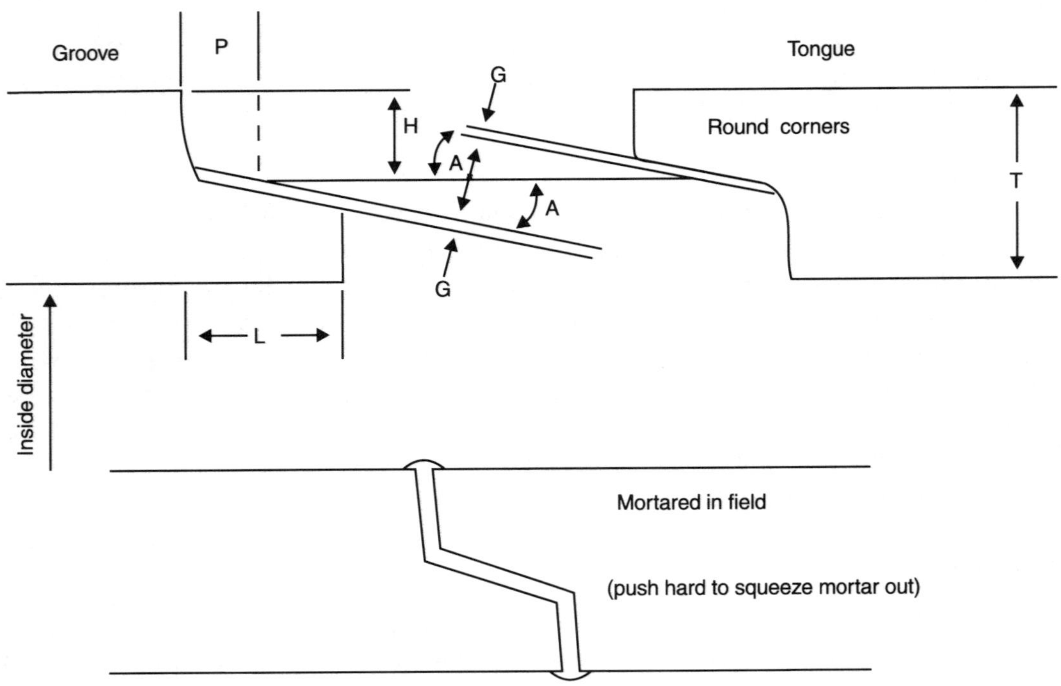

where

T = wall thickness
L = ¾ T
P = ½ L
H = ½ T
G = Varies with pipe diameter - 0.8mm for 150mm pipe to 3.2mm for 600mm pipe
A = 10 degrees

Figure 6.5. Tongue and groove design

Source: Koluvek (1970)

- A suitable solvent or mixture of solvents is applied to the surfaces. In solvent-bonding the solvent usually contains some dissolved PVC polymer or compound (commonly up to about 10%, but up to 30% in some special, strong cements). These products are known as solvent cements. When compared with straight solvents, the use of solvent cements for bonding offers the advantages of reduced evaporation loss at application, better control of the jointing and less problem with the solvent spreading outside the contact area.

  (Note that solvent cements are not gap-filling, mainly because the amount of polymer they contain is proportionately low. For good results it is important that the surfaces of the pipes fit closely together with no gaps).

- The joint is left open for a short time, while the solvent is absorbed into the surfaces. The pipe surface will seem to swell a little. American Society for Testing of Materials (ASTM) specifications indicate this should be no more than about 30 seconds, and less in hot conditions.

- The pipe and joint are brought together to close the joint under positive pressure. Because in the jointing of pipes it is not practical to apply pressure perpendicular to the joint it is particularly important that the initial fit between the surfaces is gap-free.

It is best to hold the joint in place for about one minute, otherwise the pipe could back out of the joint.

Points to remember while making a joint

Cutting the pipe

The pipe should be cut square with a fine-tooth saw. The burrs at the end of the pipe must be removed and chamfered so that when inserting the pipe the softened pipe surface is not removed. To enhance the penetration of the solvent, the surface of the pipe should be slightly roughened, using emery paper or wire wool.

Test joint for fit before gluing

The fit should be snug if the pipe and fitting have been correctly manufactured. The pipe should become tight before it reaches its full depth in the fitting. If the pipe goes in too easily a little extra solvent cement can help, but if the problem continues to recur it should be taken up with the manufacturer.

Considerations affecting jointing

Generally you will need to work quickly to complete the jointing before the solvent cement is too dry. It is preferable not to joint pipe in very high temperatures, for example when the pipe or air temperature is above 45°C. In colder temperatures you will need to wait a little longer for the solvent to penetrate the pipe surface, typically two to three minutes. Scraping the area of a test piece with a knife will show you how far the solvent has penetrated.

Applying primer and cement

In some jointing procedures a separate primer or solvent is used to dissolve the surface, before applying the cement which contains some of the dissolved polymer.

Avoid applying excess cement as this can weaken the pipe and fitting.

When forcing the pipe all the way into the socket, using the mark on the pipe as a guide, you should twist the pipe a quarter turn as this helps to distribute the solvent cement. If sand or grit is trapped the twist helps to seal the grit into the joint.

The minimum overlap for the pipes being joined should be half the inside diameter of the pipe or 75mm, whichever is greater.

Table 6.1 **Recommended minimum curing times for solvent cement joints**

Air temperature °C	*Curing time (hours)*
15 to 40	0.5
5 to 15	1
-5 to 5	2
-15 to -5	4

Source: ASTM

Wipe any excess cement from around the pipe at the socket, to stop the solvent from continuing to dissolve into the pipe and weaken the assembly.

Curing time for the solvent-cemented joint

A short time (generally 30 minutes) after the joint is complete, the pipe and joint can be handled if done so gently and carefully. The joint does not reach its full strength, which can be equivalent to the pipe material, until nearly eight days after jointing, depending on the temperature. Strength is usually adequate for a pressure test 24 hours after jointing.

ASTM recommendations for minimum curing times are given in Table 6.1.

In practice the joint reaches maximum shear strength at a steady rate over time, as shown in Table 6.2 for a particular type of PVC. Different materials may harden at slightly different rates. After one day the pipe has achieved over 70% of its final strength. Note that this final strength will depend on the quality of the jointing procedure. If the joint is stressed early on this will permanently weaken the joint.

The main advantage of solvent cement bonding is its speed and reliability. The pipe can also be joined outside the trench and subsequently lowered into place. The solvent cement is however expensive and figures quoted for consumption per joint vary widely.

Typical solvent cement consumption is in the range 0.9 to 1 l/m^2 of pipe surface or 40ml for a 150mm joint (MMP, 1989d).

Thin-walled uPVC pipe products

For jointing the newly developed corrugated and spirally wound uPVC pipe materials, a number of techniques are used. In some cases uPVC joiners are used with solvent cement. These fittings include internal tapered joiners, standard rigid uPVC pipe fittings and hand-moulded uPVC fittings. Some fittings use rubber O-rings but these are not generally designed to withstand any significant pressure.

For joining the spirally wound uPVC pipe to itself, a solvent adhesive based on polyurethane has been specified by at least one manufacturer. For hand-moulding of fittings for spirally wound pipe, a number of compounds have been used, typically comprising a suitable solvent with a filler such as glass fibres in order to caulk the joint.

Asbestos cement pipe

Asbestos cement pipe for low-pressure applications in drainage and sewerage utilizes an external collar joint with internal recesses containing rubber O-rings. Spigot socket pipe is also available with rubber gaskets achieving the water seal.

Collar jointing of plain-ended pipe

For plain-ended low-pressure pipe, the most commonly used fittings are asbestos cement sleeves.

Table 6.2 **Summary of joint strength development**

Time from assembly (hours)	*Percentage of final sheer strength*
0.5	19
1	20
4	37
8	48
16	67
32	83
48	84
96	97

Source: Titow (1985)

Because defects or scoring on the joint surface will affect the watertightness of the joint, pipes and joints should be checked before lowering them into the trench. A rasp can be used to smooth any small faults but larger faults may require the pipe to be cut back and re-turned.

Good jointing will depend on all parts being clean, the pipes being in line and level, proper location of the joint sleeve (collar) and carefully reading and following the manufacturers' instructions.

The joint and pipe surfaces must be thoroughly lubricated, and the jointing should be possible without excessive force. If significant resistance is encountered then withdraw the pipe and repeat, checking alignment.

Small changes of direction can be accommodated but should only be made after a pipe has been jointed correctly in the straight, and the trench should be wide enough for this to be done. Where a vertical curve is being negotiated, the pipe will have to be supported while jointing in the straight before being lowered. After jointing, the free pipe end can then be moved out of line by an amount usually specified by the manufacturer. Suggested limits for UK-manufactured pipe and joint systems are listed in Table 6.3.

Spigot socket pipe

Care is necessary to avoid damaging the socket ends of the pipe during installation. Although some repair using mesh and cement mortar is possible to pipes which are only lightly damaged, additional loose collars are usually necessary to replace broken collars.

As with non-reinforced concrete pipe, recommendations regarding the direction of laying of spigot socket pipe are somewhat conflicting.

There is some evidence from pressure tests that pipelines laid with sockets facing upstream are less likely to leak at the joints. This was thought to be because the opening between the socket and pipe wall did not function as an air trap. Where conditions are difficult for pipe-laying and leakage results from the difficulty of this pipe orientation then placing the sockets at the downstream end may facilitate jointing (Plusje, 1981).

On the few systems identified using asbestos cement pipe, plain-ended or spigot socket pipe was used with rubber gaskets providing the seal.

Table 6.3 **Bending limits for asbestos cement pipe**

Nominal dia. (mm)	*Offset per pipe length* 3m	4m	5m	*Angle of deflection*
75 to 100	260mm	350mm	-	5°
150 to 300	-	280mm	-	4°
350 to 500	-	210mm	260mm	3°
600 to 750	-	140mm	180mm	2°

Source: Everite (1987)

Junctions and outlets

Pipe junctions will be required in the following circumstances:

- at a riser and outlet;
- at an air vent or riser of diameter the same as or less than the pipe;
- at a branch in the pipe system.

The junction can be achieved in one of three ways:

- by use of a larger diameter stand into which the various pipes are joined;
- by use of a ready-fabricated fitting;
- by the fabrication of a fitting on-site.

In general, non-reinforced concrete pipe uses on-site fabricated fittings, while uPVC and asbestos cement pipe use factory-made fittings.

Non-reinforced concrete pipe systems

Although available the cost and difficulties of supply mean that prefabricated fittings are unpopular and generally unnecessary. Three different methods of constructing the junctions on site are commonly used. These are:

- mortar jointing of two sections of plain-ended pipe;
- casting the pipe tee using a mould;
- casting a junction block into which plain-ended pipes are fitted.

Mortar jointing plain-ended pipe

This method involves chipping a "V" section out of the barrel of one pipe. A second pipe with the end chipped to a matching shape is then positioned in the notch. One advantage of this junction is that it will allow pipes in the same horizontal plane to be joined at different angles. This method is illustrated and summarized in Figure 6.6.

Openings should be full size, within 25mm of the diameter of the connecting pipe, but overcutting will make it difficult to achieve a watertight joint. The join is then wrapped with a strip of sacking (or jute) soaked in cement slurry, and a layer of mortar is then applied over this (see the description of the jute and mortar band joint). In order for a tee or riser to withstand the load of an alfalfa valve outlet or other above-ground structure and to prevent leakage, the joint is frequently encased in concrete. Without this mass concrete

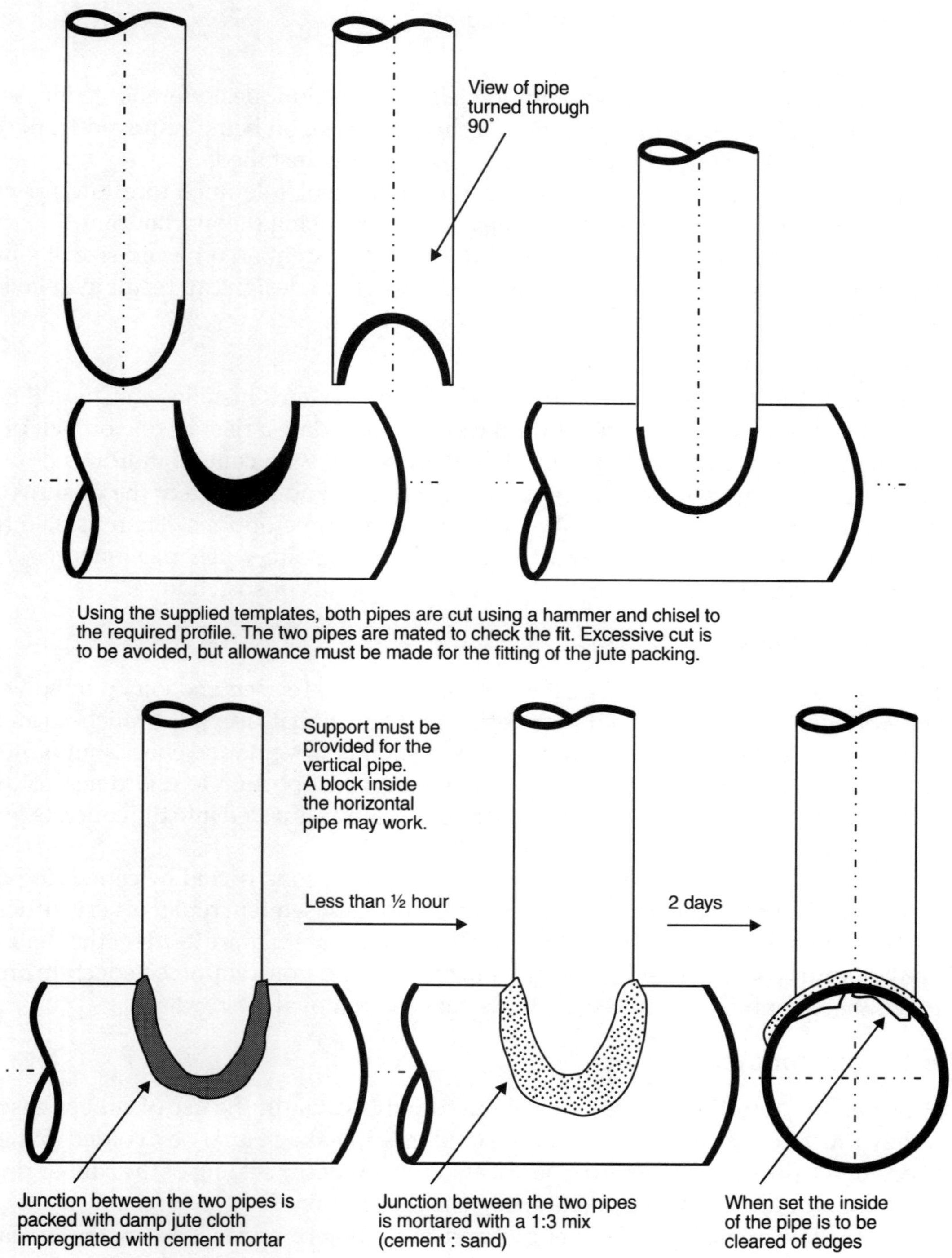

Figure 6.6. The making of concrete pipe junctions

Source: MMP (1989a)

surround these joints have a tendency to leak, though the surround adds considerably to the cost of the joint. The tees can still be major sources of leakage in the system where careful construction and testing are not practised.

Casting the tee using a mould
This involves on-site replication of a commercially-used technique. Reference to its use is found in IDP WRC, 1990. Although many of the tees broke on being removed from the mould, those which remained intact did not leak on being installed.

The junctions need to be manufactured with sufficient tolerance to allow for the variation in pipe diameter. Successful sealing is possible using the mortar joint.

One problem with casting joints is the need to plan the number, type and size of joints required before they are required. Breakage and changes to design can result in delays if adequate stocks of joints are not available.

Casting a junction block
This alternative junction is manufactured and used on systems installed in China (DST, 1990). The block is made from concrete and can accommodate a riser or one branch pipe line. The pipes fit into recesses in the block and are sealed with cement mortar.

This junction is only able to joint pipes at right angles and because of the quantity of material required it may be the most expensive of the three options. There is also the problem of portering the block from the manufacturing site when machinery is not available. The block does allow for standardisation in joint construction.

uPVC pipe systems

In almost all cases standard prefabricated tees and fittings are used and joined by solvent cement welding. This allows rapid installation and additional risers or branches can be retrospectively fitted with ease. In some structures such as air vents and check stands, non-reinforced concrete pipe can be used for the above-ground portion to guard against the problems of vandalism. The uPVC pipe is then connected and sealed into the concrete pipe using cement mortar.

Although outlet risers on the uPVC pipe system can be constructed by cementing the alfalfa valve directly into the pipe riser using a coupling, subsequent repair is very difficult if the valve or riser is damaged. In many cases alternative materials are used for the shallow and above-ground portions, including galvanized iron and non-reinforced concrete pipe. Use of a flanged joint or concrete seal allows for easy removal for repair.

Asbestos cement pipe systems

Prefabricated tees and fittings are normally used but because of the use of rubber gasket joints they must be manufactured to high standards if leakage is to be avoided. Where variations in the dimensions of the prefabricated fittings occur sealing of the rubber rings can be poor and grouting with cement mortar will be required.

Fittings may be prefabricated either by hand-moulding or by machine methods. Care must be taken when using the hand-moulded fittings as they can be weaker than the pipe and prone to breakage.

Anchor blocks or thrust blocks

For horizontal or vertical changes in pipeline gradient of more than 15° anchor blocks are necessary to resist the thrust forces. They transfer the forces to the surrounding soil. Anchor blocks prevent pipe movement leading to disruption of the joints and subsequent leakage.

Anchor blocks are required in the following locations:

- all horizontal and vertical bends greater than 15°;
- changes in pipe diameter (reducers);
- tees and pipe junctions;
- where there is an in-line valve (gate, pressure or air relief);
- at the end of the pipeline, i.e. endcaps and plugs.

They can be constructed of:

- concrete poured to fill the space between the pipe and undisturbed trench wall;
- soil of a coarse texture with cement at a ratio of 1:12 of cement : soil, which is thoroughly compacted.

In some cases compacted backfill can be used instead of an anchor block (on the outside of a radial pipe bend). The backfill should be compacted to a density equivalent to that of the undisturbed soil.

The magnitude of thrust forces to be withstood can be calculated using two simple formulae:

- For blank ends and tee junctions

$$T = 10.2 \times A \times P$$

- For axial thrusts at bends

$$T = 10.2 \times A \times P \times 2 \text{ Sin } (a/2)$$

where T = thrust force (kN)
A = external cross-sectional area of the pipe (m^2)
P = maximum internal pressure (metres of water)
a = angle of deviation of the bend

In general thrust forces due to pipe deviations of about 15° or less can be provided for by careful sidefilling of the pipe, and do not require special thrust blocks.

At each position the total thrust must be transmitted to the bearing surface over sufficient area to provide the necessary reaction force. Typical safe bearing pressures are given below. The bearing area required for support is simply the total thrust divided by the safe bearing capacity of the soil.

Required thrust block area = $D \times L = T/B$

where D = depth of the thrust block (m)
L = length of the thrust block normal to the thrust (m)
B = safe bearing pressure (kPa)

Note kN/m^2 = kPa

The bearing area will be the outside diameter of the pipe (depth) times the length of the surface normal to the thrust.

Standards such as ASAE S261.7 (1989) specify that the depth of the thrust block should be a minimum of 150mm or equal to the external diameter of the pipe, whichever is greater. The anchor block should be at least the full diameter of the widest part of the pipe. The area of the anchor block which is in contact with the soil should be a minimum of $0.5m^2$ or as calculated.

Safe bearing pressures (kPa) for some typical soils are listed below. There is a wide variation in the definition of the soil texture and the representative bearing capacity so it is best to be conservative, in the absence of actual data.

Typical safe bearing strengths (kPa):

Sound shale	500
Cemented gravel and sand	200
Coarse and fine compact sand	150
Medium clay	100
Soft clay	50
Muck	0

Source: ASAE S376.1 (1989)

Reducers and branches

Where the thrust is normal to the trench wall, the required support area of concrete thrust blocks at pipe line ends or branches is the difference between the cross-sectional area of the block and the trench cross-section. Blocking may not be required for cross junctions where the branch lines are long, that is more than 100m.

Bends

Straps are cast into the anchor blocks to secure pipelines at major changes in vertical direction. For concrete pipes, galvanized steel straps are suitable while with uPVC pipe a more flexible rubber-based material is preferred. Where concrete is used with plastic pipe or fittings, they should be wrapped in heavy plastic sheet or equivalent compressible material to prevent point stresses developing.

ASAE S261.7 (1989) and ASAE S376.1 (1989) detail recommended anchor block arrangements and the method of calculating the minimum area of the thrust block. Typical anchor block layouts are shown in Figure 6.7.

Curing and backfilling

All pipe needs to be uniformly and continuously supported over its length on firm stable material. ASAE standard S376.1 states that special attention should be given to backfilling and bedding procedures for low-pressure pipes of 450mm and larger diameter. Special design and analysis may be required to establish the strength of the material to be used as a backfill in the case of these larger diameter pipes. The standard also requires that low-pressure pipe be water-strutted or water-filled before backfilling. This is not practical in many cases but with uPVC pipe will help to keep pipe shape during backfilling and avoid floating of the pipe.

Non-reinforced concrete pipe

Curing of the pipe joints is essential if watertight joints are to be achieved. Poorly cured joints crack and leak. They can be kept moist before backfilling by covering with wet sacking (or jute bags) or simply covering with damp soil. During installation pipe openings should be sealed off to prevent dry air circulating within the pipeline. After backfilling, the dry soil over the joints can be moistened from time to time to assist joint curing.

Backfilling should begin as soon as the pipe joints have cured sufficiently so as not to be easily damaged and should be within 12 hours of the joint being constructed. The joint can be protected from damage during backfilling by wrapping a piece of plastic or heavy paper around the mortar band at jointing.

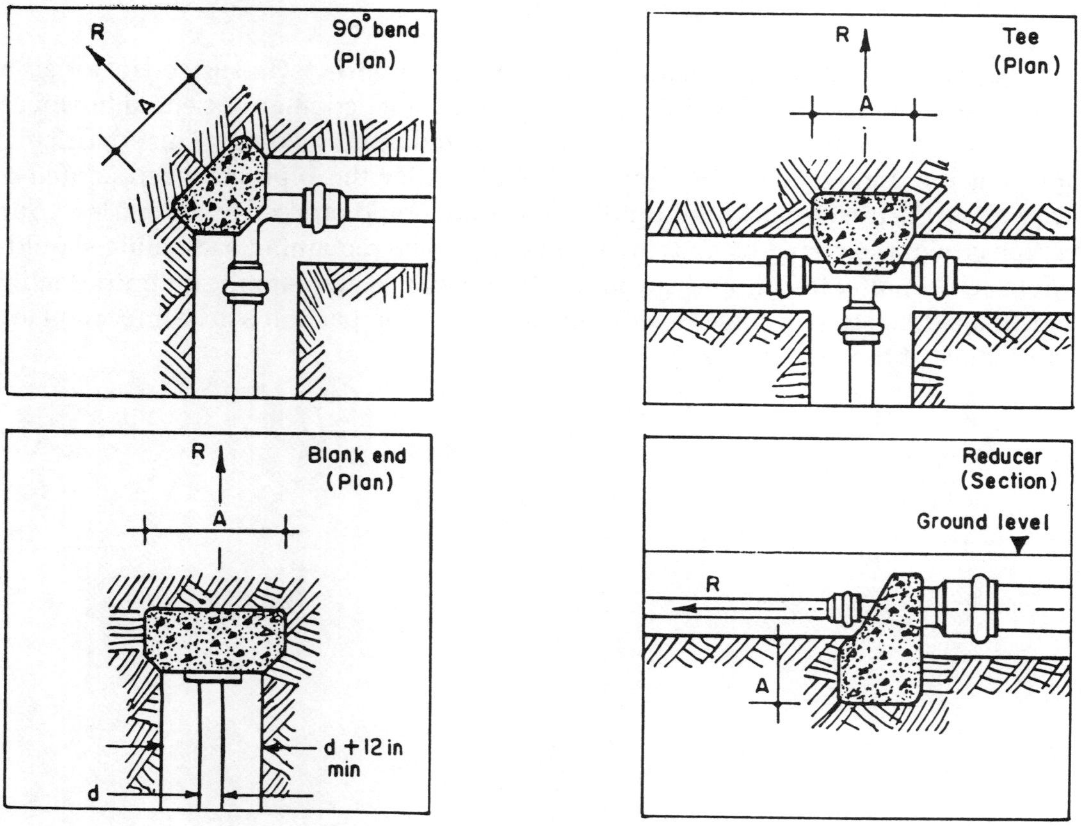

In all cases the required area of the thrust block at position A is obtained by dividing the resultant force R by the safe bearing capacity of the soil.

Figure 6.7. Examples of anchor blocks

Source: MMI (1992)

Moist soil free from hard clods, stones and excessive organic matter should be used to backfill the trench, all the time carefully compacting it around the pipe. Using moist soil minimizes the moisture stresses in the pipe by evening out the moisture content of the pipe shell. The trench need only be initially backfilled from 100 to 300mm above the crown of pipe. This will protect the pipe from large expansion and contraction forces resulting from diurnal variations in temperature. These stresses can rupture the joints and pipes, and may result in telescoping of pipes or pipes being pushed into structures.

The trench should be left open until testing has been carried out and any necessary repairs effected. Care should be taken to avoid filling the trench with water as this could cause floating of the pipe with resulting disruption of joints and alignment (see Chapter 7). The trench can be backfilled completely when repairs have sufficiently cured and the retest has been completed.

uPVC

For rigid plain-walled uPVC pipe at cover depths up to 1.5m most in-situ soil materials will be satisfactory as backfill. Careful backfilling and compaction are essential however to provide the lateral support for the pipe to carry soil and traffic loadings. This is particularly important for thin-walled uPVC pipe materials. After the pipe has been jointed and lowered into the trench, backfilling and compaction should proceed in 100mm layers until the pipe crown is covered by 150 to 300mm of soil. The remaining backfilling should be completed only after testing and any necessary repairs. Where significant traffic loadings are envisaged, selected backfill material and adequate compaction will be important to the performance of the pipe under load.

CHAPTER 7

POST-CONSTRUCTION ISSUES

Inspection and testing

The aim in construction should be to ensure through adequate standards and training that the work is of the highest quality. The testing of samples of the pipe at the point of manufacture is essential to avoid later problems and must include pipe strength (three edge bearing test) and safe hydrostatic pressure. A typical test rig for three edge bearing tests is illustrated in Figure 7.1. A simple arrangement for hydrostatic pipe testing is shown in Figure 7.2. Illustrations of both sets of testing equipment are given in the publication 'Photographs of Buried Pipelines for Surface Irrigation'.

Close inspection of all work is neither possible nor desirable, except in the pilot phase where training of staff is a major objective. Strategies which assist with improving the quality of construction include:

- involving the irrigators in supervising work done by contract labour and encouraging them to take responsibility for ensuring that their system performs as well as it can;
- employing local people to construct their own systems;
- withholding a portion of the contract payment as bond until the pipe system has been successfully tested and repaired;
- retaining trained staff to work on future schemes where possible.

The inspection and testing should include the following stages.

Visual inspection

One person should be responsible for visually checking each section of the pipeline before it is backfilled as potential problems can often be identified at this early stage.

Hydraulic test during construction

Random tests of short sections of completed pipe during construction can assist with identifying problems at an early stage. This will frequently consist of simply filling the pipeline with water and identifying leaks.

Final hydraulic test

All pipelines in the system should be tested to the full working pressure when they have cured sufficiently. Although complete backfilling should be avoided until final testing is finished this is not always possible. In almost all soils, water leaking from the pipeline will be visible at the surface from all but the smallest leaks. However in coarse-textured soils the pipeline must be only lightly backfilled to assist leak detection. In all cases sufficient backfill must be placed to prevent movement of the pipe.

In most cases the pressure test comprises filling of the pipeline to its working pressure. A test rig used to test uPVC pipe systems in Indonesia is shown in Figure 7.3.

Post-construction maintenance

Monitoring should continue for several weeks of operation of the system to identify further problems as the pipeline achieves full strength.

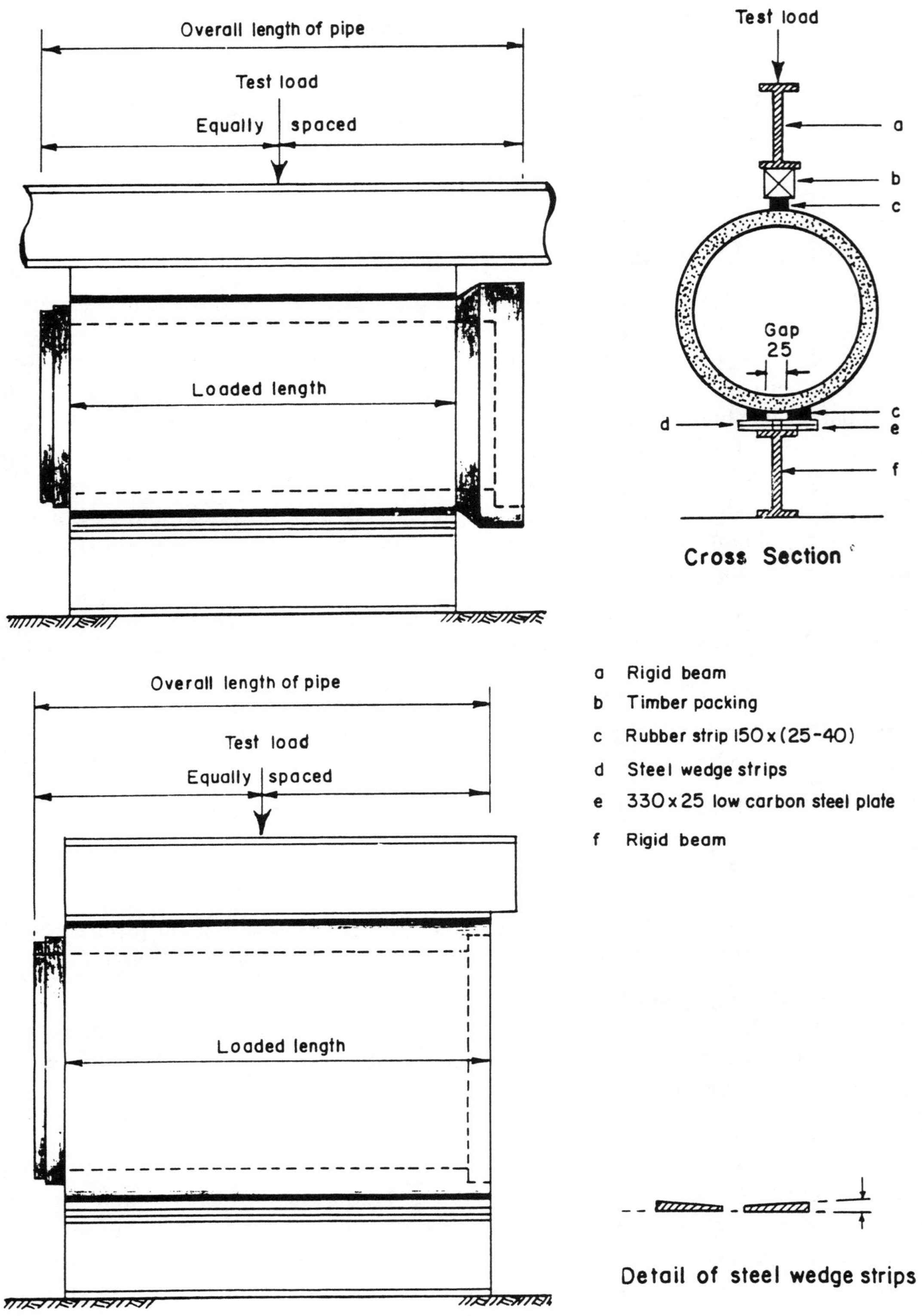

Figure 7.1. Testing arrangements for three edge bearing test

Source: MMI, 1992

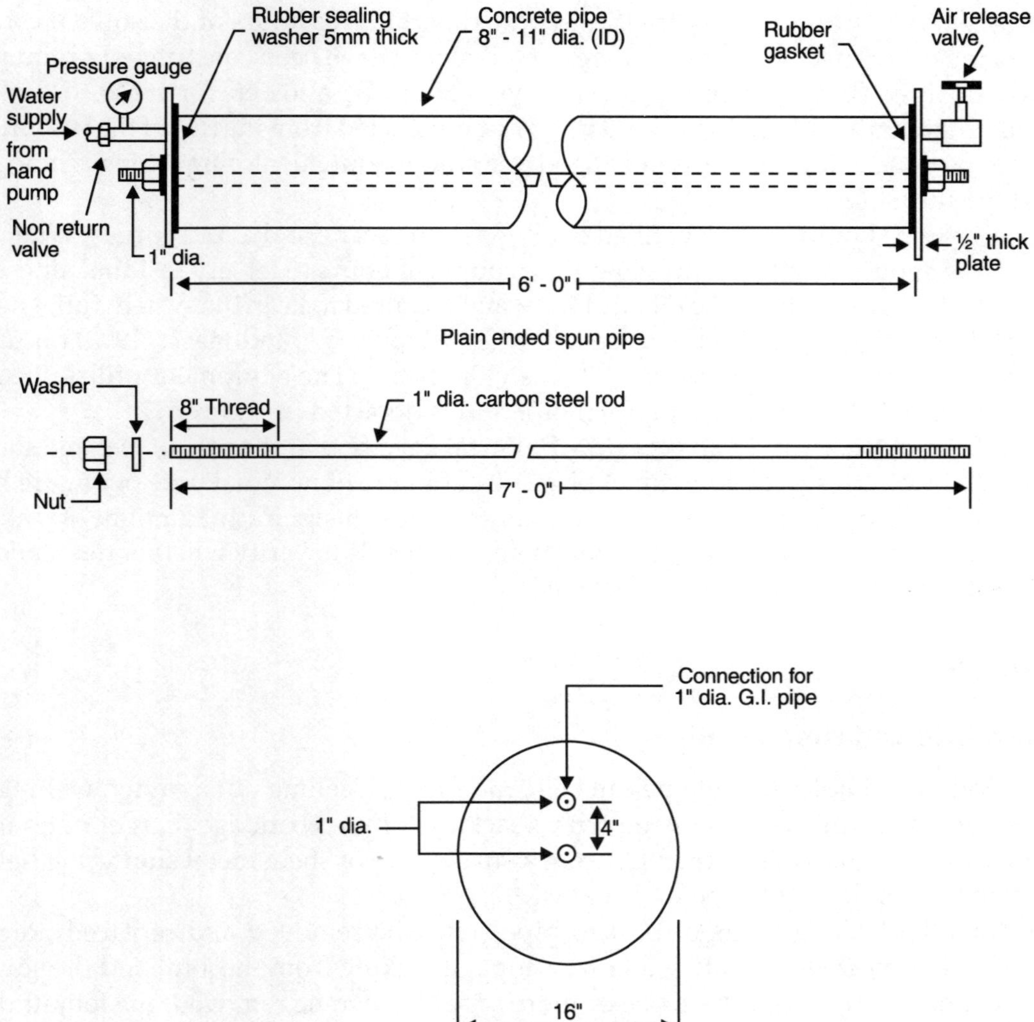

Figure 7.2. Hydrostatic pressure testing equipment

Source: MMI, 1992

Non-reinforced concrete pipe

Before hydraulic testing of any pipeline section, a period of at least seven days must be allowed for joint curing. For pipe tests during construction, tests should be done with a water level of at least 1m above the ground surface and higher if possible.

Once the entire system has been completed and the joints adequately cured the system can be tested to its working pressure. Speed of installation is important to minimize the disruption to agricultural activities and persuade farmers to wait until testing is carried out before completing backfilling.

uPVC

In theory PVC pipes can be hydraulically tested within five hours of the solvent cement joint being completed. However, where anchor blocks have been constructed a minimum period of three days (preferably seven days) should be allowed for these to develop adequate strength. The section to be tested must be isolated from the rest of the system and this can be done using a blank end flanged adaptor or equivalent plug which is braced to resist the thrust forces.

The system should be slowly filled to allow air to escape and avoid surges. When full the system should be allowed to stand for a couple of hours to let any residual air escape and then measurements can be taken of the water required to keep the system full. The use of portable testing equipment has been reported (Republic of Indonesia, 1990) for uPVC pipe systems up to a test pressure of 2 bars (20m head). The equipment utilizes a small sprayer pump belt driven by a petrol motor and is depicted in Figure 7.3.

Sir M. MacDonald and Partners (MMP, 1989a) specify that the rate of leakage should not exceed two litres per kilometre of pipeline per metre of nominal bore per metre head of test pressure per 24 hours. This represents an almost insignificant amount of leakage. No results from existing buried pipe systems are available to verify whether this standard has been satisfied for the systems installed.

Repairs

Non-reinforced concrete pipe

Small cracks and holes in the pipe can be repaired by chiselling out a cavity, wetting the surface and applying a mortar band with a sacking (or jute) bandage. Larger holes up to 50mm can be similarly repaired but using a small piece of sheet metal shuttering held in place on the inside of the pipe by wire (MMP, 1989a).

Substantial leaking joints or broken pipes must be removed and replaced. Repairs should not be carried out until water has stopped leaking from the joint and the soil has had a chance to dry out. Where one or more pipes have to be removed, the longitudinal compression forces set up during pipe-laying are partly lost as the pipes move closer together. This can be partially prevented if the line is allowed to dry out as much as possible (for several days to a week) before repairs are made. To assist the drying out, all water should be removed from the line (by pumping, or natural percolation and drainage) and all valves, stands and gates opened to facilitate air circulation. Sections of pipe used in making repairs should be thoroughly dry when installed and again moist soil should be used for backfilling.

For troublesome joints or junctions, mass concrete can be used to encase the joint but this is likely to be expensive for more than a few joints on each system.

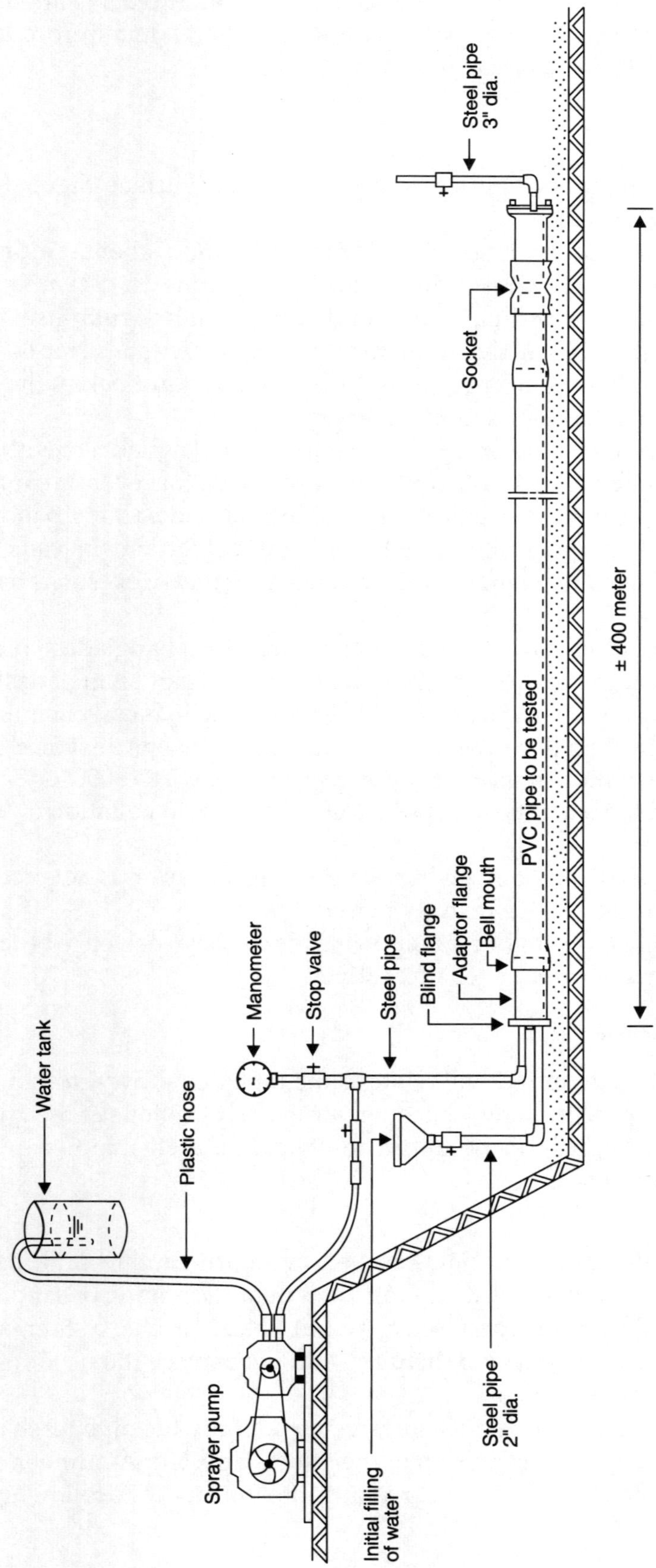

Figure 7.3. Hydrostatic pressure test equipment

Source: Republic of Indonesia, 1990

uPVC

Leaking joints should be cut out and a new section and joint fitted. No attempt should be made to re-use the old spigot socket joint. Jain Ribloc (1991) detail procedures for joint repair of their thin-walled spirally wound uPVC pipe.

Maintenance

One of the major advantages of buried pipe systems over surface channels is that they require very little maintenance.

Most of the maintenance cost is associated with the repair of above-ground structures. This is often due to damage caused by misuse or unauthorized tampering with for example, valves, rather than ordinary wear and tear through regular use. In the case of concrete pipe, installation below the depth to which expansive soils crack and expand in response to changes in temperature and moisture content (and below the soil-freezing depth) is important to avoid high maintenance costs.

Problems have been encountered on TADP schemes in Bangladesh, with rigid mortar joints on plain-ended pipe. Leaks have continued to develop for several years after scheme completion. In comparison to the collar joint, the mortar and sacking bandage joint has little strength with the pipe in tension, and it is thought that settlement leads to disruption of the joint. Although inconvenient the pipe repairs are quickly executed by local labour at a relatively low cost.

Rashid et al (1992) estimates the cost of repairing a leak on Bangladesh pipe systems at Tk 50 (Tk 55 = £1). The number of leaks reported in a study of eight Bangladesh buried pipe systems varied widely but ranged from 0 to 10, representing in most schemes less than 1% of the entire operating expenses of the scheme. Only in one instance is very high expenditure on repair of the pipe system reported, which was Tk 18000 for East Kutubpur, where persistent problems had been experienced after poor installation of pipe of mixed quality (Mridha, 1993).

Repair of outlet valves and damaged above-ground structures is not recorded for any of the schemes during the two-year study period.

Typical maintenance costs for buried pipe systems can be expected to be low compared with lined and unlined open channel alternatives.

Maintenance methods

Most of the maintenance required will relate to the repair of above-ground structures as well as the repair of pipe leaks which become particularly obvious at the ground surface during the dry season. In addition there are a number of different maintenance techniques which merit further discussion.

Pipe blockage

On occasion pipe systems will become blocked by debris or silt buildup. Clearly it is preferable to avoid this occurring by providing adequate screening or sand trap facilities where necessary. Children unsupervised may put objects in the system, and although screens on pipe stands and vents will help they do not replace the need for community responsibility and education.

Where a pipe block is suspected the aim is first to identify the pipe reach involved and this should be possible from observation of the outlets affected. A number of techniques can be tried which avoid the necessity of breaking into the pipe. The following devices can be used:

- A small portable high-pressure pump is used to supply a high-pressure hose which is progressively fed down the length of the line between two valves. The flexible high-pressure hose (25 to 40mm diameter) with specially shaped radial nozzles or rotary brush would be inserted at each delivery valve and the discharged water and sand pumped from the neighbouring valve (Cunningham, 1986). The Uttar Pradesh Tube-well Project II suggested the use of this technique for any of their systems which require cleaning.
- A double disc made of two circular pieces of wood held about 30cm apart by bolts is dragged through the pipe by means of a rope attached to it. This arrangement can be useful for removal of silt deposits from small pipes.
- Wooden or rubber balls called "beach balls" or "pills" are tied with a rope and allowed to flow down the pipe. The ball dams up the water behind it until sufficient pressure is created when the water forces its way with a scouring velocity between the ball and the invert of the pipe. The ball is of slightly lesser diameter than the pipe.

System life

Typically, solvent-welded uPVC pipe is quoted as having a life of up to 50 years. Good quality concrete pipe can have a similar life span though under the same conditions it may be slightly shorter. Pipe life will depend very much on the quality of pipe manufacture, particularly for concrete pipe and the care taken during installation. Likewise thin-walled uPVC will be much more sensitive to damage during handling and installation than standard uPVC pipe of 2.5 or 4 bar.

Field engineers quote a life span of 20 to 30 years for large diameter cast-in-place concrete pipe with only minor maintenance required in the first 15 years (Bealey, 1987).

Aggressive soil conditions

Aggressive acidic soil or groundwater conditions seldom occur naturally and even then the acid in contact with the concrete pipe is neutralized and results in a neutral zone around the pipe. Continued attack will only occur when there is a rapid acidic groundwater flow. This is only likely in some landfill sites and mining situations.

Of more concern are sulphates found in alkaline soils such as in the dry western states of the USA and some parts of the Middle East. Sulphates in solution must penetrate the concrete and be concentrated by evaporation to cause damage. The use of sulphate-resistant cement (Type II in the USA) is recommended to make cast-in-place concrete more sulphate-resistant (Bealey, 1987).

Freezing can cause problems in some areas. However the high strength, low water-cement ratio concrete of quality precast concrete pipe has excellent resistance to freeze-thaw forces. The problem arises not so much from damage to the pipe as from the ground heave which can occur with soil-freezing conditions and result in disruption of the joints and pipe alignment. If excavation below the freezing depth is not practical, insulating materials can be used. Alternatively the pipe should be drained and this will require the provision of drain valves at convenient low spots where the line can be evacuated by gravity or pumping.

Higher flow velocities also contribute to the resistance of the pipe to freezing. In freezing conditions water must not be left in the pipeline when it is not in use for any more than six to eight hours.

CHAPTER 8

COST ANALYSIS OF BURIED PIPE DISTRIBUTION SYSTEMS

Introduction

In this chapter outline costs for different types of buried pipe systems are presented and compared with open channel alternatives. Only data where both buried pipe systems and open channel systems are detailed are discussed in full.

Most of the system cost information presented and discussed is for systems in south Asia, and was collected either during field visits or is drawn from the collaborative research work conducted in Bangladesh. Because of the range of system designs used in the different projects described, very specific conclusions on comparative costs are not possible; however, general trends and relationships can be established.

Both the data and the methods of analysis are described in detail, to enable others to adapt these to their own situation.

Pipe system costs are compared on the basis of their unit length as well as with regard to the overall system costs. System costs are firstly compared with the inclusion of the value of land lost and cost of water control structures provided (for open channels) and secondly compared on a per hectare basis taking into account the effect of improved transit efficiency on the command area.

Apart from capital costs other components of cost examined include the value of land taken by the distribution system, maintenance and repair costs, and expenditure on fuel for pumped systems and on labour in all systems.

General comparison of alternative distribution system costs

Data sources

In the following Tables 8.1 to 8.6, cost data for systems in Bangladesh, India, Indonesia and Nepal are presented. They are discussed in the sections which follow. While most of the data is drawn from Bangladesh and Nepal sources, some very general costs are given for other south Asian situations. No data were obtained for systems using the newer thin-walled uPVC pipe materials now available.

Each country and region will clearly have slightly different requirements for the design of their buried pipe system. When examining cost information from specific countries it is important to recognize that unique conditions will influence the choice and cost of buried pipe technology.

Specific features which are important in the choice and cost of buried pipe systems in Bangladesh include:

- The absence of local sources of aggregate in many parts of the country necessitates the use of locally manufactured brick broken into chips as aggregate for all types of concrete.

- Because of the high level of unemployment and underemployment, labour wage rates tend to be very low, making material cost savings more valuable than labour savings.
- The high intensity of land settlement, and the highly fragmented land ownership pattern, not only results in high land values, but is also an obstacle to the establishment of new improved irrigation distribution systems.
- The absence of local petrochemical resources and manufacturing facilities, and the shortages of foreign currency, have led the government to impose high rates of tax both on the resin raw material and finished uPVC products including irrigation pipes, to encourage the use of indigenously sourced materials.

Unit capital costs

Unit cost data are drawn largely from Bangladesh and Nepal sources, and are supported by data from a study prepared for the Indonesian government. Because the pipeline component of the buried pipe system comprises from 60 to 90% of the cost of the total system, unit length pipe or channel costs can be used to compare different pipe systems, particularly where system designs are similar. For example Table 8.5 gives the cost of uPVC pipe in the pipe system built by IDA DTW II Project in Bangladesh as 84% of the cost of the distribution system.

The unit capital costs of open channels exclude the costs of necessary water control structures, land take by the open channels or infrastructure associated with the pipe system. These costs are included in the comparison of system capital costs detailed below.

Bangladesh

Bangladesh data for three different projects are detailed under the heading of construction costs in Tables 8.1, 8.2, 8.3. The data indicates the unit construction costs of earth channel systems to be some 10 to 15% of the cost of providing non-reinforced concrete pipes.

Comparable concrete and brick lining systems are considerably more expensive than concrete pipes, because of the higher material content of brick aggregate and cement. At best (concrete precast) lining alternatives are only 20% more expensive, while at worst (totally brick-lined channels) they can be up to 300% of the cost of concrete pipe. Because of the high level of government tax and tariffs mentioned, uPVC pipe materials are considerably more expensive than to concrete pipe materials particularly as pipe size increases. While data from Table 8.3 indicates that uPVC pipe systems are more than twice

Table 8.1 Comparative costs of distribution systems, Tangail, Bangladesh (1988)

Present value of costs in Tk per metre of canal/pipe

System	*Land*	*Construction*	*Maintenance*	*Total (PV)*
Earth	33.3	16	26.7	76
BPD	nil	130	8.7	139
Brick-lined	26.7	340	high	400c
In-situ concrete	26.7	130	high	200c

Source: Gisselquist, 1989

the cost per unit length of concrete pipe systems, data from Table 8.2 indicates the costs per unit length as comparable between concrete and uPVC materials, where the uPVC pipes costed were of small diameter (150mm).

Table 8.2 Comparative costs of distribution systems, BIADP, Bangladesh

Distribution system	*Cost in Tk per (m)*	*Right of way (m)*	*Land cost (Tk/m)*	*Total cost (Tk/m)*
Earth channel	29.5	3.05	64	93.5
Pucca brick lining	710	2.13	45	755
Precast semicircular	240	1.5	32.1	272.1
In-situ semicircular	356	1.5	32.1	388.1
Low-pressure concrete pipe 10" (250mm)	225			225
6" (150mm) dia. uPVC pipe	256			256

Note: Land value = Tk210 000 per ha (US$6000 per ha).
Source: Prokashali Sangsad, 1987

Table 8.3 Comparative costs of distribution systems, Tangail, Bangladesh (1990)

System	*Right of way (m)*	*Present value of costs as at 1989-90 Tk/m*		
		Land	*Construction*	*Total*
Unimproved earth channel	2.4	59	39	98
Improved (compacted) earth channel with control structures	3	75	105	180
Brick lining	3	75	584	659
Precast semicircular	1.5	38	558	596
In-situ semicircular cement concrete	1.5	38	558	596
Cement concrete buried pipe (10"/250mm)	-	-	308	308
uPVC pipe (8"/200mm)	-	-	770	770

Note: Land value = Tk 250 000 per ha (US$7140) per ha).
Source: Rashid et al, 1992

Table 8.4 Suitability comparison of various types of open channel lining with pipe systems

Lining/pipe channel system *Flow = 35 l/s*	*Suitability score*	*Cost/m (Rp x 1000)*	*Ratio of pipe cost*	*Life expectancy without major rehabilitation*
Lined channel alternatives				
1. Concrete, in-situ non-reinforced trapezoidal thickness 50mm	50	28.4	1.33	20
2. As for No. 1 with mesh reinforcement	55	32.4	1.51	25
3. Concrete precast, segment unreinforced, 350 bar	65	27.9	1.30	25
4. As for No. 3 reinforced 300 bar	65	31.9	1.49	30
5. Concrete, precast slabs reinforced, 225 bar	37	28.0	1.30	10
6. Brickwork, double layer, pointed	34	24.2	1.13	10
7. Brickwork, double layer, plastic layer, top layer only pointed	44	25.1	1.17	10
8. Stone masonry 1:4 mortar trapezoidal 200mm thick	37	31.2	1.46	10
9. Ferrocement, in-situ, 35mm with mesh reinforcing trapezoidal	52	26.2	1.22	15
Pipe system				
10. uPVC pipe 50m head 1m deep	65	21.4	1.00	25

Source: Republic of Indonesia, 1990

Indonesia

The notable feature of the Indonesian data presented in Table 8.4 is the very competitive cost of uPVC pipe systems compared with other hard surface lining materials, suggesting either indigenous sources of uPVC resin for manufacture, and/or low rates of government taxes and tariffs.

Table 8.5 Cost summary for buried pipe systems in India and Bangladesh

Item	*Uttar Pradesh (uPVC loop) (Rs)*	*US$*	*TADP (concrete) (conc. branch) (Tk)*	*US$*	*Faridpur (conc. branch) (Tk)*	*US$*	*IDA DTW II (uPVC loop) (Tk)*	*US$*
1. Tower or tank	19 550	850	2 000	57	6 382	182	30 800	880
2. Cost of pipe			114 175	3 262	122 058	3 487	451 972	12 913
3. Valves			9 835	281	13 876	396		
4. Jointing			19 920	570	33 352	953		
5. Trenching			14 363	410	29 950	856	40 937	1 169
6. Other structures			27 708	791	15 530	444	7 748	221
7. Field channels	9 750	424	10 500	300	10 500	300	10 500	300
Sub-totals								
Dist. chamber	19 550	850	2 000	57	6 382	182	30 800	880
Dist. system	216 200	9400	186 000	5 315	214 766	6 136	500 657	14 304
Field channels	9 750	424	10 500	300	105 000	300	10 500	300
Total	**245 500**	**10 674**	**198 500**	**6 300**	**236 071**	**6 618**	**541 957**	**15 483**
Command (ha)		100		40		40		40
Cost per ha		1 067		158		165		387

Note: conc. = non-reinforced concrete pipe
Source: Unpublished material from TADP and IDA DTW II Projects Bangladesh, and Uttar Pradesh Tube-well Project II, Lucknow, India

Despite the wide range of lined channel alternatives examined, the differences in unit cost are small (10-20%), with typically the stone masonry and reinforced concrete channel segments proving the most expensive, probably because of the greater quantity of higher cost materials used.

Nepal

Scheme cost estimates for open channel distribution systems, prepared by GDC (1987), are based on the average of unit rates collected in the Terai area of Nepal, and are given in Table 8.6. Costs for buried pipe system components are based on rates for pipes, fitting and alfalfa valves at the pipe outlets, for comparable systems built in Uttar Pradesh, India.

Scheme estimates for the capital cost of lined and unlined channels set these at 178% and 14% respectively of the cost of uPVC pipes, not including the provision of control structures and the value of the land take. The channel lining in this case is half brick and relatively expensive compared to prefabricated concrete-lined alternatives used in other parts of the world.

Table 8.6 **Summary of deep tube-well distribution system costs, Nepal**

		300m³/h			*150m³/h*	
Item	*Piped (uPVC) (Loop)*	*Lined (Brick)*	*Unlined (Earth)*	*Piped (uPVC) (Loop)*	*Lined (Brick)*	*Unlined (Earth)*
Well						
Borehole	373.6	373.6	373.6	232.2	232.2	232.2
Pump	112.0	112.0	112.0	80.2	80.2	80.2
Electrification	300.0	300.0	300.0	172.5	172.5	172.5
Sub-total	785.6	785.6	785.6	484.9	484.9	484.9
Wellhead works						
Pumphouse	24.4	24.4	24.4	22.4	22.4	22.4
Discharge box	0.0	7.6	7.6	0.0	5.7	5.7
Control tank	76.5	0.0	0.0	51.0	0.0	0.0
Land	5.0	5.0	5.0	5.0	5.0	5.0
Sub-total	105.9	37.0	37.0	78.4	33.1	33.1
Total well costs	891.5	822.6	822.6	563.3	518.0	518.0
Distribution						
Channels	6.2	1148.0	89.6	3.1	496.0	46.2
Buried pipes	636.5	0.0	0.0	318.2	0.0	-
Structures	0.0	123.4	110.5	0.0	70.9	66.7
Land	0.0	22.5	22.2	0.0	9.9	8.7
Sub-total	642.7	1293.9	222.3	321.3	576.8	121.6
Totals						
Construction	1534.2	2116.5	1004.9	884.6	1094.8	639.6
Engineering (10%)	153.4	211.7	104.5	88.5	109.5	63.9
Overall total	1687.6	2328.2	1149.4	973.1	1204.3	703.5
Command area (ha)	88	80	60	44	40	30
Costs per hectare		(Rs'000)				
Distribution system*	7.3	16.2	3.7	7.3	14.4	4.1
Total unit	19.1	29.1	19.2	22.6	30.1	23.5

Note: *including 10% for engineering.
All costs in Nepalese Rupees.
Source: GDC (1987)

System capital costs

Distribution system cost including structures

While unit cost comparisons give an initial indication of the relative cost of buried pipe systems, they underestimate the cost of lined and unlined channel alternatives, by not including the cost of providing water control structures or the value of land lost.

Bangladesh

When the cost of structures within the distribution system and the value of land lost are added, Rashid et al (1992) suggest that earth channel systems cost in the order of 55 to 60% of a concrete pipe system (refer to Table 8.3). The improved earth channel has a right of way, which at 3m is twice the width assumed for precast concrete lining and the value of the land occupied is estimated at 40% of the cost of the distribution system. Note that improved earth channel systems with water control structures are considered to be about twice the cost of unimproved earth channel systems.

Data from Gisselquist (1986a) suggest earth channel systems to be about 40% of the cost of concrete pipe systems (refer to Table 8.1) where the difference in maintenance cost is not included. This is also supported by data from Prokashali Sangsad (1987), detailed in Table 8.2.

Overall pipe system costs presented in Table 8.5 indicate the substantial cost difference between concrete and uPVC pipe systems built in Bangladesh, with the uPVC pipe system being more than twice the cost.

India

Although no detailed data have been obtained for any particular system, general system costs were noted for pipe systems built using a range of designs and materials in central and northern India. In Table 8.5, costs for the uPVC pipe system built in Uttar Pradesh are compared with costs for concrete pipe systems in Bangladesh (TADP and Faridpur) and a uPVC pipe system built by the IDA DTW II Project.

When considered in terms of US$ per ha, the Indian uPVC pipe system is 1.7 times the cost of the concrete pipe systems built in Bangladesh, but still considerably cheaper than the uPVC pipe system built in Bangladesh, because of the different tax and tariff structures which apply. The Bangladesh pipe systems serve about 40 ha while the Indian systems serve 100 ha.

Other pipe systems in India for which overall costs were identified included the following:

Scheme name	Rs per ha	US$ per ha	
1. Vindhyasini akkadselarge lift irrigation scheme	16 560	920	(1986)
2. Pipaldagarhi lift irrigation scheme	29 200	1622	(1990)
3. Gadigaltar tank irrigation scheme, total scheme	40 350	1613	(1991)
Gadigaltar tank irrigation scheme, distribution system	9770	390	

Source: Data collected during research study trip (van Bentum, 1992)

Nepal

When the cost of land taken by the distribution system, and the provision of structures, is included the capital cost of lined and unlined channels rises to 198% and 35% respectively of the uPVC pipe systems considered (refer to data in Table 8.6). The cost of the land take is considered to represent only 10-15% of the total cost of providing a system with earth channels.

USA

While the USA possesses the largest area currently served by buried pipe distribution systems, very little relevant cost information has been obtained for these systems. Keller (1990) summarizes data on USA costs for a range of modern irrigation systems among which buried pipe systems for surface irrigation can be considered one option. These detail the cost of precision surface irrigation (precise land levelling) with buried pipe systems as ranging from US$ 800 to 2500 per ha. This is more expensive than low-cost sprinkler application techniques but cheaper than solid set sprinkler systems and localized irrigation by drip or spray, which cost in the range US$ 2000 - 3500 per ha.

He also concludes that costs for similar systems in developing countries will be from 25 to 100% higher because of the small farm sizes, freight costs, foreign exchange and technical difficulties.

The upper cost range for buried pipe systems refers to surface irrigation where large fields are precisely levelled by machine. If the small basin irrigation systems with pipe systems found in developing countries are considered to be at the lower end of the cost range, then expected costs without adjustment would be in the vicinity of US$ 1000 - 1200 per ha. Only two Indian schemes (costs already summarized) have system costs which approach these. Bangladesh system costs, already mentioned, are a small fraction of these, probably because of the use of a high proportion of local labour and materials, and an emphasis on system affordability.

System costs including improved transit efficiency

When the overall system costs are considered in terms of the cost per hectare of land irrigated, taking into account the higher transit efficiency of pipe systems, a different picture emerges. Table 8.6 provides summary data for comparative system costs in a Nepal context while Table 8.7 provides summary data based on Bangladesh experience, and must be considered purely indicative of likely costs.

The two analyses compare capital costs for the distribution system as well as the entire surface irrigation system. The water supply in both the Nepal and Bangladesh systems comprises a deep tube-well. Costs do not include operation and maintenance expenditure, or any allowance for depreciation and eventual system replacement.

Transit efficiency

In order to compare per hectare costs for buried pipe systems with both lined and earth channel alternatives, representative transit efficiencies for each system must be estimated.

Table 8.8 gives transit efficiency estimates based on Bangladesh field work, and other data. Average pipeline transit efficiences of 80% for concrete pipeline systems and 90% for uPVC pipe systems are considered realistic, while earth channels and lined channels are estimated to have 50% and 70% transit efficiencies respectively. This is supported by research work conducted in Bangladesh which suggested that concrete pipe systems had

Table 8.7 Capital cost comparison for Bangladesh distribution system (US$'000)

Item	*Non-reinforced concrete pipe*	*uPVC*	*Earth*	*Brick*
Well, pump, motor and pumphouse	17.1-22.9	17.1-22.9	17.1-22.9	17.1-22.9
Discharge box	0.0	0.0	0.2-0.3	0.2-0.3
Header tank	0.4-0.6	0.4-0.6	0.0	0.0
Sub-total Well costs	17.5-23.5	17.5-23.5	17.3-23.2	17.3-23.2
Distribution channels	0.0	0.0	1.4-1.7	14.3-20.0
Buried pipes	8.6-11.4	15.7-18.6	0.0	0.0
Structures	included	1.1-1.7	1.1-1.7	1.1-1.7
Sub-total Distribution system	8.6-11.4	16.8-20.3	2.5-3.4	15.4-21.7
Overall total	26.1-34.9	34.3-43.8	19.8-26.6	32.7-44.9

Source: van Bentum (1992)

at worst 25% of the losses typical on lined channel systems and less than 10% of those on earthen channel systems (Rashid et al 1992). PVC pipe system losses are assumed to total 10% based on an average rate of outlet leakage on 40% of the outlets in the system (observations of systems in Uttar Pradesh, India; van Bentum, 1992).

Bangladesh

The irrigation system distributes water from a deep tube-well with a pump discharge of 56 l/s. Four possible irrigation distribution systems are considered, including an earth channel system with water control structures, a brick-lined system, a non-reinforced concrete pipe system and a uPVC pipe system.

The following assumptions have been made regarding the design and operation of the schemes costed:

- The concrete buried pipe system has a total length of 1800m comprising three separate 600m branches.
- The uPVC pipe system comprises two loops of 150mm pipe, each of which is 1300m in length.
- The earth channel system is 2000m in length and of compacted earth and includes concrete water control and outlet structures.
- The brick-lined channel is also 2000m in length and has outlet and water control structures.

The costs relate to a typical Bangladesh command area with relatively flat topography and few obstructions. Where the command is undulating or dissected to any extent the open channel distribution systems are likely to become significantly more expensive if not infeasible in some situations.

Field channel systems are considered equivalent in extent and area for each of the schemes for simplicity, although in practice differences may well exist.

The cost analysis indicates that earth channel systems are about 30 to 35% (distribution system cost only) and 70 to 80% (where well costs are included) of the cost of concrete pipe

Table 8.8 **Field measurements of transit efficiency**

Pipe material	*Seepage lps/100m*	*Pipe length (m)*	*Discharge (lps)*	*Transit efficiency (%)*
1. Non-reinforced plain-ended concrete pipe with sacking-wrapped joint				
a. average	0.33	500	35	95
b. poor	0.69	800	40	86
2. uPVC	NK	1500	56	95
3. uPVC	NK	300	20	95
4. Earth channel	7.7	3-400	45	50

NK: not known
Source: Rashid et al (1990) and van Bentum (1992)

distribution systems. In this case compaction is assumed and structures for water distribution are provided.

Brick-lined channel systems are twice the cost of concrete pipe systems, with uPVC pipe systems comparable in cost to brick lining at both the distribution system and total system (including well costs) levels.

System costs per hectare of land irrigated

Ignoring non-engineering factors, the irrigable area for a particular discharge can be said to depend on the transit (pipeline distribution) efficiency. The effects of this are illustrated by the following analysis, based on a maximum net irrigable area of 50 hectares with a transit efficiency of 100%. Poorer transit efficiencies will result in smaller command areas and higher capital costs per hectare of irrigated land as shown in Table 8.9.

The relative costs (which are independent of the 50 ha assumption) show clearly that the more efficient buried pipe systems should give a lower capital cost per hectare than an open channel system, as the fixed cost of the well is spread over a larger area.

When considering the distribution system alone, earth channel system costs rise to nearly 50% of the cost of concrete pipe alternatives, while uPVC pipe systems are 1.7 times and brick lining more than two times the cost of concrete pipe systems.

Where total well costs are included (deep tube-well) earth channel systems become comparable (US$924 per ha), and even slightly more expensive than concrete (US$707 per ha) and uPVC pipe systems (US$815 per ha), while brick lining remains the most expensive alternative (US$1104 per ha).

Similar calculations would apply in other situations, for example, if the irrigable area is fixed the source could be smaller and cheaper (whether a tube-well, canal or reservoir) with a buried pipe distribution system than with open channels, though savings may not be as significant.

Nepal comparison of system cost

Data from Nepal used in the second cost comparison, and detailed in Table 8.6, are based on the buried pipe systems designed and constructed in Uttar Pradesh, India and described in detail by van Bentum (1992). Systems of this type with elevated tanks and loop uPVC pipelines have also been constructed on the Terai area of Nepal, under the Bhairawa Lumbini project (GDC, 1987).

The earth and lined channel systems are assumed to comprise 1300m of channel with associated water control structures, so a level of water management is possible similar to that available with buried pipe systems. Two different sizes of irrigation system were examined by GDC (1987), to see whether economics of scale applied. In a similar way to the Bangladesh model the differences in transit efficiency are assumed to result in a larger irrigable command area for a fixed inflow discharge.

The relative transit efficiency estimates chosen by GDC (1987) are very conservative, in the light of the tenfold difference in seepage rates, measured between unlined channels and pipeline systems during Bangladesh-based fieldwork (Mridha, 1993). Pipe systems are considered to be 90% efficient while lined and earth open channels are assumed to have transit efficiencies of 80% and 60% respectively.

If considering the distribution system costs over a fixed command, an earth channel distribution system is equivalent to 60% of the cost of a buried pipe system, while a lined system is more than twice the pipe system cost of NRs 7300/ha. When the command area is varied in relation to the transit efficiency achieved, earth channels become equivalent in cost to buried pipe systems at around NRs 20 000 per ha, while lined channel systems remain 50% more expensive at NRs 30 000 per ha.

Lower operation and maintenance costs are likely to further increase the gap between buried pipe and lined channel systems.

Table 8.9 **Transit efficiency, irrigable area and costs per hectare for Bangladesh systems**

Item	*Non-reinforced concrete pipe*	*uPVC pipe channel*	*Earth*	*Brick lining*
Transit efficiency (%)	80	90	50	70
Irrigable area (ha)	40	45	25	35
Costs per ha (US$)				
distribution system	215-285	373-451	100-136	440-620
Well costs (US$)	438-588	384-522	692-928	494-663
Total costs (incl well costs)	653-873	757-973	792-1064	934-1283
Relative costs	0.86-1.15	1.00-1.28	1.05-1.41	1.24-1.69

Source: The costs are drawn from Table 8.7

The above analysis of pipe system costs in Bangladesh and Nepal assumes that transit efficiency is the critical constraint. Fieldwork carried out in Bangladesh (Rashid et al, 1992) found that irrigated areas on the selected tube-wells were considerably less than design, and were constrained principally by poor scheme management and inadequate extension services (see Figure 1.5). This is supported by evaluations of the performance of buried pipe irrigation schemes, such as have been completed by Kolavalli and Shah (1989) on the Uttar Pradesh Project and by Pant (1989) on the new design of public tube-wells in Uttar Pradesh.

Unless socio-economic, institutional and managerial issues are considered more carefully during design and resolved at an early stage, then the full potential of buried pipe systems is not likely to be realized.

Non-capital cost consideration

Apart from the cost of the land taken by open channel systems, no data have been obtained on quantified savings in non-capital costs. Based on the field evaluation of pipe systems visited during the research, a summary of relative costs for each of the different pipe systems is presented below in Table 8.10. Real costs are not presented but rather cost ratios based on the expected cost for a uPVC system expressed as "X". In all cases it is considered that pipe systems provide a considerable saving in maintenance and operating costs and operator and farmer labour.

Operating costs for fuel and oil are considered to be indirectly proportional to the transit efficiency though an allowance for the 3 to 4m increase in the delivery head on pipe systems is made by reducing the open channel operating cost ratio by 10%.

Cost of land take

The information presented in Tables 8.2 and 8.3 indicates the average right of way for earth and brick-lined channels to be 2 to 3m. If a typical open channel distribution system serving 40 ha is assumed to have 1400 to 1800 metres of channel, then the land take would range from 0.28 to 0.53 ha (0.7 to 1.3% of the command).

Work completed in Bangladesh estimates the saving of land due to the use of buried pipe systems as averaging 1.2%. Although this represents only 0.48 ha on a 40 ha command, in some intensively settled areas very high land prices can make this a valuable saving. For example in Bangladesh where land in Tangail district has a value of Tk 200 000 to 250 000 per ha, an area of 0.48 ha represents 25 and 45% of the capital cost of a uPVC and concrete pipe system respectively.

Strictly the value of the land take should be considered in terms of the income lost rather than the sale value of the land, however the comparison is illustrative and helps to assess the land's value for wealth creation and the provision of living space.

Maintenance costs

Actual data on the level of maintenance costs for buried pipe systems compared to open channel systems have not been obtained, although a number of references suggest estimates based on monitoring of costs for high pressure systems. Rouke (1984) suggests that maintenance costs for high-pressure pipe systems total 0.5% of the replacement cost of the pipe system.

Gisselquist (1986a) assumes an annual expenditure on pipe systems of 1% of the capital cost, and for an earth channel system a sum of 25%, which is included in the costs presented in Table 8.1. Despite the differences in capital costs this gives earth channels an annual expenditure twice that of buried pipe systems.

An important part of the estimate of the maintenance cost of buried pipe systems, is the expected length of service for the pipeline. As a pipe system approaches the end of its life the maintenance costs of the system can be expected to rise significantly. While the life of non-reinforced concrete pipe systems manufactured in south Asia is unclear, pipe systems built in the early 1930s and 1940s in the USA are still functioning reliably according to reports by Baudequin et al (1990) and Merriam (1987a). High-pressure uPVC pipe systems have now been in service for well over 20 years and Walton and Elzink (1988) in assessing the condition of existing pipe installations conclude that the 50-year design life currently used for uPVC pipe may be too conservative, and that under good conditions a 100-year design life is more realistic.

Labour requirements

The opportunities provided by buried pipe systems to reduce the labour involved in scheme operation and maintenance are difficult to quantify. Ways in which buried pipe systems reduce the labour input to irrigation systems are described.

- By cutting the length of open channels the labour required for annual maintenance is reduced.

- On closed and semi-closed pipe systems where operation can be controlled from the head of the system, an operator is required to spend less time directing water to the particular part of the command, provided the necessary irrigation schedules have been agreed.

- Because systems with higher transit efficiencies potentially deliver higher flow rates to the field, the farmer can spend less time irrigating and improve farm water management.

Table 8.10 **Ratios of non-capital costs**

Item	*Non-reinforced concrete*	*uPVC pipe*	*Earth channel*	*Brick lining*
Land take (ha)	nil	nil	0.25-0.5	0.2-0.4
Maintenance costs	1.5X	X	3-5X	3-8X
Operating costs fuel	1.2X	X	1.7X	1.3X
Labour (operator)	1.5X	X	3X	2X
Labour (farmer)	X	X	2X	1.5X

Source: van Bentum (1992)

Conclusions on cost data

General

With all the emphasis commonly given to channel lining, it is surprising to discover evidence that pipe systems have a comparable cost. However the very limited amount of data is not enough to prove the cost competitiveness of buried pipe systems. This should however be sufficient to give designers and irrigation practitioners the confidence to try pipe systems, if only because the operating advantages over more traditional lining solutions appear to be gained at little extra cost.

Unfortunately very little useful information has been obtained on the maintenance and operating costs of buried pipe systems.

The improvements in transit efficiency demonstrated with buried pipe systems are not always reflected in increases in the irrigated area. Spreading the cost of the water source over larger command areas and achieving more equitable flow division will only achieve increases in irrigated area where limited water supplies, well-developed organizational structures and high-value crops provide the incentive for irrigators to co-operate.

Although the cost data are limited, some general conclusions can be made regarding buried pipe systems which have been built in south Asia.

Capital cost

In all the situations examined, pipeline distribution systems can be built which are less expensive than a wide range of hard surface lining system alternatives. In general concrete pipe systems will be considerably less expensive than uPVC systems where the same level of service is provided. Whether uPVC pipe systems are competitive with open lined channel systems will depend on pricing mechanisms in-country for uPVC pipe raw materials and the types of lining system commonly built. The brick lining systems widely constructed in south Asia will generally be considerably more costly than buried pipe alternatives.

While earth channel systems will be at worst 50-60% of the cost of concrete pipe alternatives, if the cost of the water source is spread over the command area adjusted for the higher transit efficiency, then the cost differences are small. The lower operating and maintenance costs would be likely to favour the selection of the pipe system.

Clearly the most competitive pipe material option will vary with country and location but in all cases careful selection should enable buried pipe distribution systems to be built at a cost which is less than any hard surface lining alternative.

Other costs

The other cost savings, such as land take by open channel systems and reduced labour costs, will be important in different places. Where land is limited and high in value, such as in Bangladesh and northern India, land take becomes an important consideration, while labour costs are an important constraint in middle income countries such as Thailand.

Pipe materials

The cost reductions possible with locally manufactured non-reinforced concrete pipe, with simpler jointing and the thin-walled uPVC pipe materials now becoming available, may provide opportunities for extending the use of buried pipe systems, by improving the benefit/cost ratio. Future cost analysis should examine lower-cost uPVC pipe systems using the newer thin-walled corrugated materials now being marketed for low-pressure applications.

CHAPTER 9

TRENDS IN THE UPGRADING AND DEVELOPMENT OF BURIED PIPE SYSTEMS

In many developing countries reducing the cost of buried pipe systems is an important condition for the wider adoption of buried pipe systems. However in higher income areas of the world, the demand for higher water use efficiency, coupled with a drive to reduce labour inputs, has encouraged investment in technology to improve the efficiency of existing buried pipe systems. Pipe systems are perceived as a lower capital and energy cost alternative to the conversion of surface irrigation systems to sprinkler and micro-irrigation systems (Merriam, 1987c).

In parallel with this has been the development of new thin-walled corrugated plastic pipe materials for use in gravity drainage and sewerage applications, which can potentially be utilized in low pressure situations (Riblok Aust., 1990).

Buried pipe systems are seen as a medium-cost alternative for the rehabilitation of irrigation systems (Manuellan, 1988). Galand (1989) and Renault (1988), describe research and development effort in France and the USA, focusing on the development of technologies which provide surface irrigation systems and particularly existing buried pipe distribution systems with efficiencies and delivery performance similar to that of alternative sprinkler and micro-irrigation systems. Many of these technologies (the most widely adopted are described below) are as yet not well known outside their countries of development.

Technologies which have moved beyond the experimental phase include:

Gated pipe

Gated pipes, described by Booher (1974), are portable pipelines used to distribute water from low-pressure outlets to individual furrows, via small shutter-controlled orifices regularly located along the pipe and facing the furrows. The pipe is normally composed of uPVC with special flexible joints for easier handling but is on occasion made from aluminium. This technology has tended to replace the traditional earth channel and siphons, and in California is now estimated to cover some 360 000 ha (Baudequin et al, 1990).

Layflat tubing

Both Galand (1989) and Booher (1974) describe flexible butyl rubber or ultraviolet-resistant PVC tubing used as "gated pipe", which was developed in order to enable machinery access to fields after irrigation. The tubing generally has flexible hose sleeves of small diameter through which water is released. Flow control is achieved by folding and clamping the sleeve, so allowing streams to be cut back or stopped.

Cablegation or transirrigation

An automated system of outlet control has been developed in France and the USA, utilizing a cable which is attached to a plug drawn down a sloping gated or buried pipe. The progress of the plug allows flow to occur from different groups of gates or risers, with the period of flow determined by the plug speed. The technique, developed in 1980, is known as "cablegation" in the USA and "transirrigation" in France and is described in more detail by Renault (1988).

Surge irrigation

This is a technique developed for use with furrow irrigation whereby the irrigation supply is intermittently pulsed, at the outlet from the buried pipe system, to create a rapid advance of the water front down the furrow by means of a series of surges. Water is delivered to the furrow from either gated pipe or individual furrow risers, and pulsing is achieved using a pulse valve which directs flow to either of two pipelines consecutively. Galand (1989) estimates that, by 1988, pulse or surge irrigation was already being practised on more than 150 000 ha in the USA alone.

Commutative valve

Renault (1988) reports the development by Neyrtec and the Société du Canal de Provence of a simple time-activated valve which allows for the transfer of irrigation supply from one irrigation line to another on completion of the allotted period of irrigation. The extent of the use of this device is not however reported.

Large on-farm streams

Merriam (1987d) sets out the advantages of providing a larger supply flow which is available at potentially variable rates. He concludes that this would enable the setting of all the furrows or border strips in a block at one time, and allow for cutting back of the delivery flows when necessary. His contention is that this can greatly reduce the labour required and can appreciably increase uniformity and efficiency. The larger supply is achieved by having an intermediate reservoir which receives a regular inflow from the conveyance system (secondary canal) and can deliver a large, variable supply. A large-capacity pipe system is required with outlets to gated pipe. A return flow system by returning runoff to the reservoir at the upper end can help achieve higher water use efficiencies and reduce drainage problems.

While certainly worthy of consideration, such large stream systems are probably most suitable for the large farm holdings typically found in western industrialized countries, such as the USA.

Conclusion

The advantages of buried pipelines over open channels have been described in Chapter 1, and these are confirmed by the widespread adoption of pipe systems on small and medium-scale irrigation schemes, where water is pumped from the source or the land has sufficient slope. Conventional large scale canal systems can also be upgraded using buried pipe distribution systems instead of tertiary canals. This will usually require pumping from the secondary canal, and can be combined with an intermediate reservoir to provide flexibility. The pumping requirements however will be considerably less than for upgrading to trickle or sprinkler methods of irrigation. Buried pipelines thus represent an intermediate technology for irrigation development. To develop such systems effectively however requires careful design according to the selected method of operation, and attention to quality in construction.

APPENDIX 1

Pipe materials commonly used for low-pressure buried pipe distribution systems

Non-reinforced concrete pipe

Mechanized vertical moulding techniques

Koluvek (1970) describes two methods which have been and continue to be widely used in the USA, namely the tamp stick and the packer-head vertical mould methods.

With the tamp stick process an internal form is used in addition to the outer mould, and stiff low-slump concrete is added slowly while a series of rods are moved continuously inside the annular space, to compact the concrete. The interior pipe shutter or core, which defines the inside diameter of the pipe, is a cylindrical tube made of cast iron or fabricated steel. The exterior form or jacket comprises two halves which can be separated or stripped from the pipe after the form is removed from the machine and set vertically on the ground. The whole manufacturing process for one pipe takes as little as a minute. To achieve adequate curing and as an alternative to the daily application of water, in high humidity conditions newly manufactured pipes can be sealed within a plastic sheet covering.

Tamp stick

Koluvek (1970) notes that the first tamp stick machines used in the USA, which included concrete batching equipment, were not automated but had relatively low labour requirements. Current pipe factories are largely automated with much higher outputs of finished pipe. In a typical installation, pipe ranging in diameter from 200 to 350mm can be produced, to a standard which conforms to ASTM hydrostatic test pressures of 30m. Pipe is usually manufactured with tongue and groove joints and in lengths of about 1.3m.

Packer head

The alternative packer head method replaces the internal pipe form with a rotating mechanical trowel. Similar quality pipe is manufactured, though the technique can accommodate a larger range in pipe size and reduces the expenditure on pipe forms. The equipment is more complex than the tamp stick equipment, but widely used in newer more automated pipe factories in the USA (Koluvek, 1970). Georgi (1989) mentions the existence of concrete pipe manufacturers with modern packer head machines in operation in Bangladesh.

Both these techniques enable concrete of relatively low slump to be compacted to give high quality low pressure pipe. Merriam (1990) suggested costs for equipment made in the USA to manufacture this type of pipe, in the range of US$30 000 for reconditioned equipment and more than US$60 000 for new machinery (1991 prices) to manufacture pipe of 100-400mm diameter and up to 1.3m in length. This does not however include the cost of moulds and accessory equipment.

Alternative vertical mould manufacturing methods use vibrators and vibrating tables to achieve improved compaction of low-slump concrete, usually by mounting the mould on a vibrating table or within a special housing. Chinese researchers (DST, 1990) in Hebei province report the use of a small vertical pipe-making plant, which uses a modified vibrating table and lifting equipment that can be operated by two people.

Vertical moulds, hand-rammed

Georgi (1989) describes pipe manufactured by compacting hand-batched concrete using hand-held rods, inside a mould comprising both core and jacket forms. The pipe moulds can be made from different materials such as wood, iron sheets, and plastic, though fabricated steel is the most common and durable. However the only information obtained concerns the use of steel forms in Bangladesh. Georgi (1989), while describing pipe manufactured by this technique with different joint types such as socket and spigot, tongue and groove, and plain ended pipe, concluded that plain ended pipe was the easiest to manufacture. Forms for pipe incorporating alternative joint types were complicated and costly to manufacture.

Georgi (1989) describes briefly pipe manufacture with vertical shutters whereby the core, comprising several longitudinal mild steel sections, collapses into the interior of the pipe for easy removal. Burnt oil is specified for application to the formwork surfaces to prevent the concrete adhering, and compaction is achieved using rods agitated inside the mould space while concrete is poured by hand into the mould.

Gisselquist (1989), in describing early TADP pipe manufacturing activities, reports that only grouting of the entire surface of the vertically cast pipes could ensure adequate sealing of the pipe. An early decision was also made to shift from spigot-socket to tongue and groove joints, in the light of difficulties encountered by masons constructing the system.

Achieving adequate compaction appears to be the major problem with vertical pipe manufacture carried out by hand, and a high proportion of the pipes manufactured have too high avoid space and hence are not water tight (Rashid et al, 1992). In discussions with TADP engineers, the practice of lowering the water:cement ratio to ease compaction was cited as a reason for the resulting weak and porous concrete pipe. Experiences with this technique in the pilot phase of the project led the Tangail Rural Development Project in Bangladesh to abandon vertical moulds in favour of centrifugally spun pipe (Georgi, 1989).

Machine spun pipe

The manufacture of concrete pipe by machine spinning is a well developed technique particularly for non-reinforced concrete pipe, in the south Asia region (Gisselquist 1989). Most large-diameter low pressure pipe and the full size range of high pressure pipe is manufactured using this technique. In some countries, including India and Bangladesh, the technique has been modified to make non-reinforced pipe in smaller diameters. Though the process is rather slow, tongue and groove, and spigot and socket pipe can also be made in this way (Gisselquist, 1989).

In common with other pipe spinning techniques, a stiffened outer mould is spun by mounting it between two axles, one of which is turned by a belt drive from a small stationary motor. Once the mould is spinning at around 200 to 300 rpm, hand-mixed concrete is shovelled into the spinning form from either end. The centrifugal force developed ensures the retention and spread of the low-slump concrete within the mould, while a long steel bar is used to achieve an even wall thickness and smooth finish.

Good quality pipe can be produced provided attention is given to material and manufacturing quality (MMI, 1990a). Concrete batching can help avoid some of the variation that occurs with hand-mixing. On occasion pipe may be of variable quality because of shortcuts taken by plant operators for example to reduce the quantity of cement used, to use poorly graded aggregate and to increase pipe output.

In Bangladesh, India and Pakistan small private workshops exist using motor driven pipe spinning equipment to manufacture non-reinforced concrete pipe. Much work has been done to scale down the size and complexity of equipment needed to manufacture pipe by this method, so that pipe quality relies heavily on close control of process and materials.

Considerable work has been conducted in Bangladesh to establish recommendations for pipe manufacture in general and the machine spinning process in particular. The results from this work are detailed in MMI (1990a), but care should be taken before extending these results to other situations as the recommendations apply to manufacture with aggregate comprised of brick chips.

Hand spun pipe

Work in Bangladesh has also led to the development of hand spinning equipment, with the aim of enhancing village level employment (Georgi, 1989 and IDP, 1990). Although the hand spinning plant will have a lower output (20 pipes per day) than the motorized equivalent (40 to 50 pipes per day depending on the number of moulds) it is of particular interest, because of its lower capital cost and greater portability compared to the larger mechanized pipe spinning plants.

Tangail Agricultural Development Project has been using this technique successfully for over five years in Bangladesh to manufacture plain ended pipe of satisfactory quality for low pressure buried pipe distribution systems.

Private sector manufacturing of non-reinforced and reinforced concrete pipe in south Asia often provides pipe of low quality (MMP, 1989a). Areas for improvement include careful batching of materials and adequate curing, while cost reductions would be provided by eliminating the reinforcement and enabling the irrigation system users to manufacture their own pipe.

Cast-in-place pipe

The American Concrete Institute (ACI, 1980) defines cast-in-place pipe as an "underground continuous nonreinforced concrete conduit having no joints or seams except as necessitated by construction requirements." The use of cast-in-place concrete pipe has been an established technique for over 25 years. In California reference is made to its use for more than 40 years in the San Joaquin Valley (ACI, 1980). In the USA, initial hand construction methods have been replaced by mechanized techniques.

The construction of cast-in-place pipe is a relatively sophisticated technology and the pipe is particularly sensitive to stresses due to temperature and moisture gradients. Although a wide range of diameters have been built, from 300 to 3000 mm, most of the pipe constructed is of sizes greater than 600mm in diameter where the pipeline is mainly used for conveyance. The minimum pipe diameter in the USA is considered to be that which allows access for a person to carry out any necessary repairs during the lifetime of the pipeline (Univ. Cal., 1977).

More recently Chinese researchers have developed low-cost techniques for constructing cast-in-place pipe by hand and using small-scale machine methods. Although the quoted Chinese costs are low, performance data is not available (DST, 1990). It is considered that cast-in-place concrete pipe is likely to be less reliable than rigid precast concrete pipe, given the sensitivity of the technique to construction methods and soil conditions and the difficulty of making joints and junctions which are leak-free.

Design considerations

American standards (ACI, 1990) state that cast-in-place pipe is intended for a maximum head of not more than 4.6 m including the effects of pressure surge and water hammer. Necessary concrete quality is specified as having a compressive strength of 20.7 MPa. Commonly, cast-in-place pipe is used for conveyance pipelines, on low-head closed pipeline systems, with open pipe stands constructed at any water control structures or shut-off valves. The pipe is considered to be able to withstand heavy traffic loads with a minimum pipe cover of 600 mm, though it must be considered to be of lower strength than comparably installed precast concrete pipe.

Chinese researchers (DST, 1990) describe the successful manufacture of cast-in-place concrete pipe in diameters from 150 to 350mm using a pressurized plastic sleeve. Limited test data on the pipe gives safe working pressures in the range of 1-2 kg/cm^2. Chinese low pressure pipe systems are typically open pipe systems with maximum working pressures seldom exceeding 5m. Clearly, even if cast-in-place pipe can be successfully installed, it will only be useful for very low pressure situations. The major concern with the use of cast-in-place concrete pipe will be pipe life, given its susceptibility to soil movement and the quality of initial construction. In the small pipe sizes used in buried pipe systems it will be impossible to repair cracks and leaks which occur, by the internal plastering method used in the larger diameter pipes (ACI, 1990).

Specifications and recommendations concerning cast-in-place pipe as used in the USA are covered by standards produced by the American Concrete Institute (ACI, 1990).

Unplasticized polyvinylchloride

Nearly all the plastic pipe used in low pressure pipe systems to date has been unplasticized or rigid PVC. It is manufactured in a wide range of sizes though pipe for irrigation use seldom exceeds 250 mm in diameter, as manufacturing and transportation costs increase dramatically in the larger sizes. Joints are usually of the spigot and socket type with solvent cement glue although rubber ring joints (z joints) have been used in some applications.

uPVC has several advantages over concrete and asbestos-cement alternatives including:

- its easier handling and lighter weight;
- its smooth bore and therefore low friction loss;
- fewer joints and easier jointing.

Its major disadvantage is its cost. In the projects identified as using uPVC pipe, in Indonesia (Republic of Indonesia, 1990), India (Cunningham, 1986) and Thailand (HHP/HTS, 1988) the pipe which has been installed is typically that rated to 25 or 40 m (2-4 Bar) allowable working head which is well in excess of the hydrostatic pressure requirements of the system. Thicker walled or higher pressure rated pipe is chosen more for its stiffness and resistance to damage during handling and installation, than for its hydrostatic strength.

PVC pipe manufacturers have over recent years developed a range of thinner-walled uPVC pipes with equivalent dimensional properties (e.g. stiffness) to those of thicker walled rigid uPVC. To reduce the quantity of PVC raw material per lineal metre the alternative pipe products utilize a corrugated outer wall, or a lower density cavity wall. Although in general the pipes are of lower cost for the same diameter, the lower material costs are partly negated by higher manufacturing and processing costs (see Table A1.1).

uPVC pipes of this type can be categorized into three groups:

- smooth bore and corrugated external wall;
- smooth bore spirally wound corrugated pipe;
- smooth bore and outer wall, with foam or cavity fill.

These are discussed in turn below. Specifications for uPVC pipe are covered by several standards, including:

- ASAE S376.1 'Design, Installation and Performance of Underground Thermoplastic Irrigation Pipelines'. June 1988. American National Standards approved.
- British Standards 312:1972 'Code of Practice for Plastic Pipework Part 1. General Principles and Choice of Material. Part 2. uPVC Pipework for Conveyance of Liquids under Pressure'.
- ASTM F1176 'Design and Installation of Large Diameter Thermoplastic Irrigation Systems (Max Pressure 63 psi)', November 1990.

Lighter-weight uPVC pipes

Smooth bore and corrugated external wall

Plastics manufacturing companies in a number of countries have developed thin-walled uPVC pipe materials, which are smooth inside and have a corrugated external wall. Applications include sewer relining, new stormwater drains, road culverts and irrigation water conveyance pipelines. Differences between pipes on the market relate to the relative reduction in weight of uPVC material per unit length, and the methods and costs of production compared with equivalent rigid uPVC pipe (Järvenkylä, 1988).

Any assessment of the suitability of any or all of the products for low pressure pipe distribution systems requires a detailed consideration of the location of the proposed pipe system and the properties of the particular pipe material. However an indication of the cost saving which may be possible can be gained from the summary of weight reduction in Table A1.1. It should be remembered that raw material cost is but one, though in most cases the largest, element of the cost of any uPVC pipe installation. A more complete comparison can only emerge when costs of manufacture, transportation and installation are also taken into account.

Some of the currently available uPVC pipe alternatives listed by brand name include:

- Ribstruct
- Dynamit nobel
- DSA
- Ultra-rib
- Double skinned pipe (by Drossbach)
- Sonrib
- Bauku
- Permaloc

Although currently these pipe materials are largely used for drainage, sewerage and ducting purposes, reference is made to the production, testing and use of such pipe in low pressure irrigation systems in China (DST, 1990).

Smooth bore spirally wound corrugated pipe
This group of uPVC pipes are constructed by extruding uPVC into various standard profiles. The profiles, once transported to site, can be wound using portable equipment to produce finished pipe in a wide range of diameters. The fastening system uses both a mechanical push-fit groove and chemical adhesive bonding to lock the profile into its finished shape.

Although a number of companies are working on pipes of this type, information has only been sourced from one (Riblok, 1990) on the successful use of the product on low pressure piped distribution systems for surface irrigation in Spain, Australia and India (Jain, 1989 and Riblok Aust., 1990).

Smooth bore and outer wall, with foam or cavity fill
The third group of lighter weight uPVC pipes retain the smooth external and internal wall properties of rigid uPVC. The weight reduction is achieved either by introducing cavities into the pipe wall, longitudinal or concentric, or by producing a lower density "foam" like uPVC material for the wall. Several different products are currently available and they are known by the following product names:

- Bipeau (important in the French sewage pipe market)
- Wavihol
- Petzetakis.

No information on the use of the pipe materials for low pressure water distribution has been obtained.

Internally corrugated pipe
The last group of lighter weight uPVC pipes are those with both external and internal corrugated walls. Such pipes have been in use for many years in land drainage applications where the rebates have been regularly perforated to allow the entry of soil water. These pipes have the greatest weight reductions compared to rigid plain walled uPVC pipe but of course lower flow capacities compared with smooth bore pipes, because of the non-planar internal surface. Several variations of this pipe are currently being produced though their use is at present restricted to storm water drainage systems (Jain, 1989).

Where the higher friction losses are not a problem and problems of siltation of the pipe are very unlikely, internally corrugated pipes are a very attractive option for new buried pipe systems. Pipes currently in production include:

- Nordisk corrugated pipe
- Brandon and Pressigny corrugated pipe.

Asbestos cement pipe

Asbestos-cement pipe, though widely used for high pressure irrigation, municipal and industrial applications, has been little used in low pressure irrigation applications. Lower pressure pipe (maximum working pressure 12m) is manufactured for sewerage and drainage applications, typically in sizes ranging from 100 to 750mm.

Although asbestos-cement pipe is usually more expensive than concrete or uPVC alternatives, adequate quality concrete and competitively priced uPVC pipe may not always be available. While claims are made of long life and low maintenance costs for

Table A1.1 Comparison of the raw material content of uPVC pipes with rigid uPVC

Weight savings of lightweight pipes (having the same two-year stiffness)

Smooth inside	*Weight*	*Weight saving (%)*
Solid wall PVC	100	0
Bipeau	80	20
Wavihol	75	25
Rib lock (Mabo)	65	35
Ultra rib	55	45
Double skinned pipe (Drossbach)	45	55
Corrugated pipe		
Nordisk corrugated pipe	35	65
Brandon and Pressigny corrugated pipe	25	75

Note: Comparison based on approximately the same inside diameter.
Sources: Riblok (1990) and selected manufacturers' data.

asbestos cement pipe supported by many years of field experience (Univ. Cal., 1977), few buried pipe systems built using asbestos cement have been identified.

Commercially produced asbestos-cement pipe is smoother than similarly spun concrete pipe, resulting in a lower coefficient of friction and allowing smaller pipe sizes to be specified with a consequent saving in overall cost. (Hydraulics Research 1990b).

Construction considerations

Pipes typically vary in length from 2 to 4 m and are normally joined and sealed using a range of rubber gasket joints (Plusje, 1981 and Everite, 1986). These comprise an exterior asbestos-cement collar or sleeve with rubber gasket inserts sealing against each pipe.

Care needs to be taken to minimize the health risks of using asbestos-cement pipe although manufacturers are adamant that there is no evidence that water quality is affected by the pipe material, and any health risks are minor provided necessary precautions are adhered to during normal handling and installation of standard pipes and fittings. Everite (1986), manufacturers of asbestos-cement pipe, detail the use of a hand saw or a power driven disc cutter which has a continuous water feed attached to keep the dust level to a minimum when cutting asbestos cement pipe. If dust levels are significant during any part of the installation, the work area should be kept damp and a dust mask worn.

Particular specifications for the use of low pressure asbestos-cement pipe in irrigation have not been found but aspects of their use should be covered by the following standards for low pressure sewerage and drainage pipe:

- BS 3656
- IS 243
- ISO 881 (sulphate-resistant cement).

APPENDIX 2

Worked example for the design of a closed buried pipe system

Introduction

The data for this design example is drawn from an IDA DTW II system in Bangladesh. The scheme is named for the purpose of this exercise after the district (or upazila) in which it is found, i.e. Bhaluka. The system is described as it was designed, and draws on data supplied by staff from Mott MacDonald International (1989, 1990) but variations to the design have been introduced where considered appropriate. The system comprises a closed pipe system with a pumped water supply from a deep tube-well water source.

The design procedure follows the steps described in the main publication, and summarized briefly below.

Step 1. General scheme survey

Step 2. Preparation of pipeline layout

Step 3. Finalize system selection

Step 4. Hydraulic design of the pipeline

Step 5. Selection and design of remaining component structures

Step 1. General scheme survey

The starting point for the survey of the potential command area was the existing land records map. Through field checking and consultation with the farmers involved, a map was prepared. Figure A2.1 constitutes this map with the later addition of the layout information. One spot level per irrigable field was found to be adequate for design purposes, and the data is presented on the map with the decimal point representing the level position.

Social and economic assessments carried out at the same time helped to identify problem areas but supported the decision to go ahead with the distribution system.

Step 2. Preparation of pipeline layout

Target command area

The peak design irrigation duty is based on the assumption that irrigated rice (boro) in the winter dry season will be the main crop over the entire command area.

An analysis of water requirements for rice suggests a peak water requirement of 3000 m^3/ha/month for the month of March. This allows for in-field deep percolation losses of 5mm/day and a contribution from effective rainfall of about 11mm/month.

The design water duty then depends on the estimate made for distribution efficiency and the maximum period of pump operation. Table A2.1 below presents a range of design water duties for boro rice for various distribution efficiencies and maximum daily pumping periods.

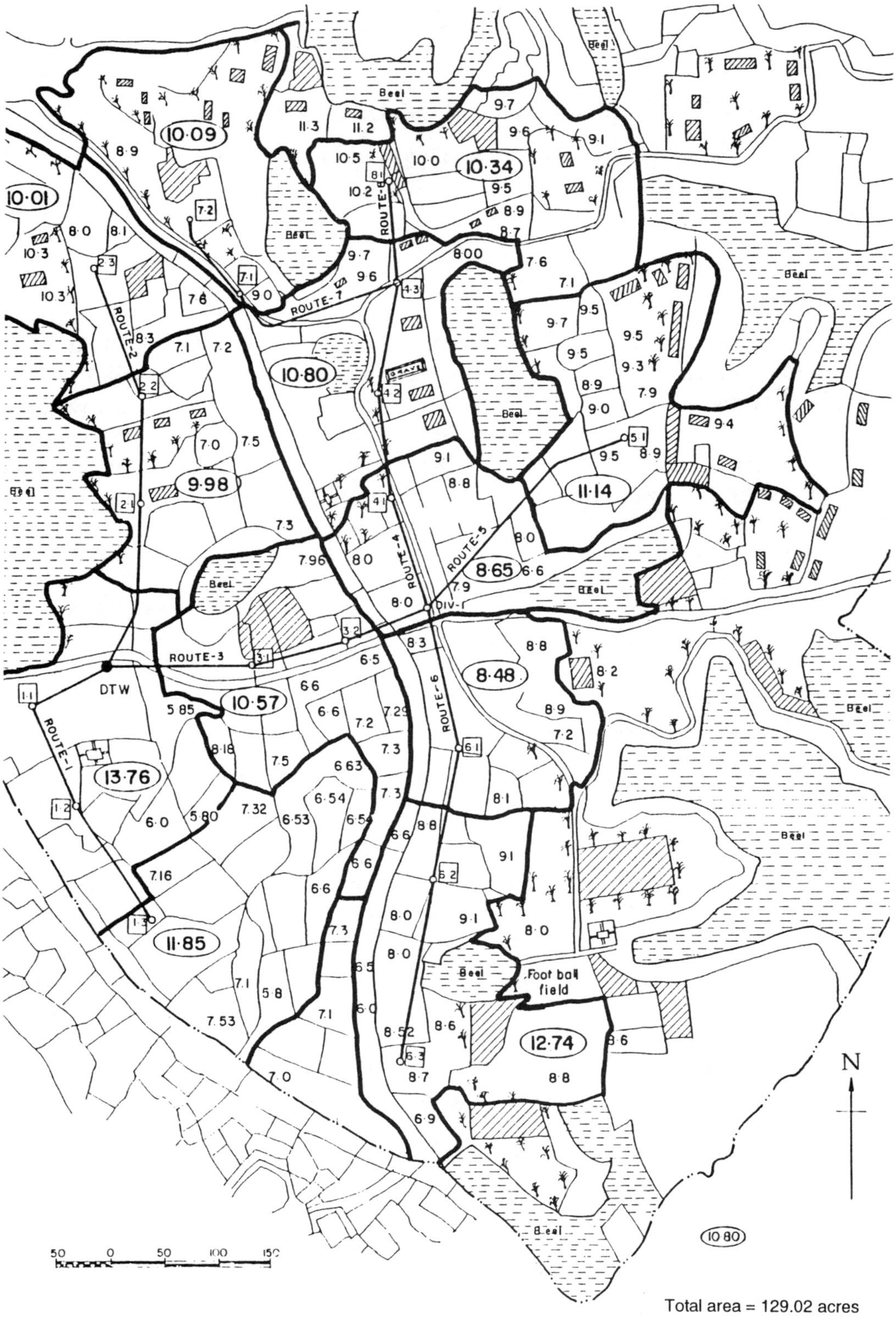

Figure A2.1. Layout and map of Bhaluka buried pipeline distribution system

Source: MMP (1989b)

Table A2.1 **Design water duty for Bhaluka (l/s/ha)**

Pumping period (hours)	*Distribution efficiencies* 60%	65%	70%	75%
12	3.86	3.56	3.3	3.08
14	3.3	3.05	2.83	2.64
16	2.89	2.67	2.48	2.31
18	2.57	2.37	2.2	2.05

The very high water requirements of rice growing during the winter dry season become clear during this analysis. The choice was made on a design duty of 1.4 l/s/ha. which at an efficiency of 65% and daily pumping, of 18 hours, would allow some 60% of the command area to be cultivated to boro rice. In practice the opportunity to improve the overall distribution efficiency, and the move to lower water demanding crops, would allow the proportion of the command irrigated to rise above 60%.

Given a design discharge from the tube-well of 56 l/s and a water duty of 1.4 l/s, the target command area was set at around 40 ha. This can be considered a lower limit as areas of up to 50 ha are supplied from one distribution system using a single deep tube-well.

Topographical features
In the case of Bhaluka the position of the tube-well was already set and there was a requirement to supply the command of the existing surface channel system. Opportunities for extending the command were most favourable on the western side of the tube-well but the community of interest and farmer preference was for the higher land on the eastern fringes of the existing tube-well command area. From the map it is clear that this land is more unevenly distributed and separated by low-lying areas which flood for a prolonged period.

Though feasible the pipe system required was more extensive than normal, and hence more expensive. In an attempt to reduce the cost, the design envisaged that the full flow would not go to the very extreme outlets but be split on the eastern side of the command at a division structure. The implications of this choice of operation became more apparent only after the scheme was completed.

Field channel lengths
In Bhaluka, as with most of the tube-well systems in Asia, earth channels are used to distribute water to individual farmer fields. Depending on soil texture the seepage losses are typically in the range of 5-10 l/s/100m where a delivery stream size of 20-30 l/s is provided at the outlet. Field channel lengths in excess of 100m should be avoided but have been accepted in Bhaluka for the outlets on the boundary of the command. Field channel slopes tend to be very flat, and short field channel lengths help to keep the depth of flow to a minimum.

Outlet discharge
Achieving high water use efficiencies in basin irrigation relies on the speed of coverage of the basin. It is desirable to have the highest flow which is manageable. Typically the farmers and their field channel systems are unable to handle more than around 30 l/s

without significant losses from spilling and leakage from the earth channels. Hence a tube-well discharge of 40-56 l/s needs to be divided between several field channels for optimal operation.

Construction quality

The leakage losses from a unit length of field channel are comparatively high because of their poor compaction. The field channels, except where crossing non-cultivated land, are generally reconstructed on an annual basis prior to the irrigation season.

Outlet command

The size of the outlet command in the case of Bhaluka has been largely established by the acceptable field channel length, given a maximum command of 4 ha. An outlet with land equally distributed around it for a distance of 100m would have a command of about 3.2 ha.

Pipeline layout and length

As already mentioned the layout of the Bhaluka scheme is very largely determined by the existing position of the deep tube-well. The well had been positioned according to other objectives including proximity to the house of the major land owner.

A branching pipe layout was chosen for Bhaluka both because of the very fragmented nature of the command area and the use of concrete pipe. From the header tank at the well there are three pipe branches. The third and most extensive branch is then again divided into three at a division stand. This allows parts of the system which are not in use to be closed off, so reducing the losses from pipeline seepage.

Step 3. Finalize system selection

The initial selection of the operating and control system, was for a header tank with bottom valve control. The design would allow for the full pump discharge to be delivered to the division structure at which point it would be split into two of the branches.

Thus the pump discharge was to be divided between two outlets, one on each of two pipe branches either at the header tank or at the division structure. With bottom valve control in both structures the precise division of flow would depend on the choice of valves and branches which would be operated. On those outlets capable of receiving the full pump discharge a outlet distribution structure with flow division would be provided.

The difficulty with this system of operation is that the precise division of flow will be determined by the particular combination of valves which are operated and their settings. In order to provide the flexibility of a supply where flow was evenly divided between any two of the pipe branches, then top water level control would be required at both the header tank and the division structure.

Step 4. Hydraulic design of the pipeline

Under the operating and control system outlined the critical path for hydraulic design will probably be where full pump discharge is delivered to the division structure and the flow is split between line 6 and line 4. The situation where valves are open at the extreme ends of either line will probably represent the greatest friction losses. The particular outlets which are critical will depend on the design outlet water levels calculated.

Available delivery head

The available delivery head is set by the maximum practical height of the pressure tower. In this case for the tower constructed from non-reinforced concrete pipe rings this is set at 4m.

Outlet design water level
Outlet design water level equals:

Head loss through the outlet
+
Head loss through the outlet distribution structure
+
Head loss in field distribution system to critical field
+
Head loss from the distribution system to the field
+
Depth of water ponded in the field
+
Ground level of the critical field.

Outlet head loss
For Bhaluka the head loss through the outlet, based on the use of a 250mm valve on a riser of 275mm diameter, will be no greater than 0.1m where the delivery flow does not exceed 56 l/s.

Head loss through the distribution structure
The maximum head loss through the structure, where "pucca nuccas" are used to divide flow and a weir to measure the outlet discharge, will be 0.2m.

Maximum depth ponded
The maximum depth of water ponded equals 0.1m.

Head loss in the distribution system
The head loss in the earth channel distribution system is defined by the length of the longest field channel in the command, given that the design slope of the field channel is 1 in 2000. The head loss from the channel into the field is not considered significant and can be ignored, given that the earth channel walls are cut to direct the water to the field.

$$\text{Head loss} = \frac{\text{Channel length (m)}}{2000}$$

Thus for a field channel length of 100m the slope will be 0.05m.
Thus the outlet water level can be defined by:

Ground level of critical field + 0.4m + Earth channel slope

The design water levels for the outlets in Bhaluka are summarized in Table A2.2 overleaf.

Minor loss coefficients
In this situation the minor friction losses of interest for the critical path include:

- the number of straight tees;
- the number and angle of the bends;
- the inlet valve head loss;
- the outlet tee head loss.

In practice the largest head loss is represented by the inlet valve loss.

Two operating regimes are analysed to establish the critical pathways for the pipe network. They are the operation of the following combination of outlets:

- 7.1 and 6.3.
- 8.1 and 5.1.

For each of these sections of the pipeline the fittings and coefficients are listed below.

Outlet 7.1 to division structure

2 × 45° bends	2 × 0.15	=	0.3
1 × 90° bend	1 × 0.2	=	0.2
3 × 22° bends	3 × 0.1	=	0.3
3 straight tees	3 × 0.1	=	0.3
1 riser	1 × 1.5	=	1.5
1 inlet valve	1 × 4.5	=	4.5
	Total		**7.1**

Outlet 6.3 to division structure

1 riser	1 × 1.5	=	1.5
1 inlet valve	1 × 4.5	=	4.5
2 straight tees	2 × 0.1	=	0.2
	Total		**6.2**

Outlet 8.1 to division

3 × 22° bends	3 × 0.1	=	0.3
3 straight tees	3 × 0.1	=	0.3
1 riser	1 × 1.5	=	1.5
1 inlet valve	1 × 4.5	=	4.5
	Total		**6.6**

Table A2.2 **Outlet water levels for Bhaluka**

Outlet	*Highest ground level (m)*	*Field channel length (m)*	*Design water level (m)*
Pump	9.2	0	9.6
1.1	8.56	40	8.98
1.2	7.88	0	8.28
1.3	8.0	0	8.4
2.1	8.62	0	8.92
2.2	9.48	0	9.88
2.3	10.3	50	10.72
3.1	8.9	0	9.3
3.2	7.96	100	8.41
4.1	9.1	75	9.53
4.2	9.48	0	9.88
4.3	9.7	50	10.4
5.1	9.7	150	10.18
6.1	8.9	100	9.35
6.2	9.1	75	9.53
6.3	8.8	150	9.28
7.1	10.3	0	10.7
8.1	10.5	50	10.93

Table A2.3 **Hydraulic calculations for Bhaluka**

Outlet	Pipe reach	Length (m)	Diameter (mm)	Flow (l/s)	Velocity (m/s)	Velocity Head	K is the sum of all the form loss coefficients	Fittings loss	Headloss (m/m)	Pipeloss (m)	Total head loss	Downstream water level (m)	Upstream water level (m)
7.1+6.3	7.1 to Div	720	275	30	0.50	0.013	7.1	0.092	0.0012	0.86	0.95	10.70	11.65
	6.3 to Div	430	275	30	0.50	0.013	6.2	0.081	0.0012	0.52	0.60	9.24	9.84
	Div to DTW	340	275	60	1.00	0.05	5.9	0.295	0.0046	1.56	1.86	11.65	13.51
8.1+5.1	8.1 to Div	440	275	30	0.50	0.013	6.6	0.086	0.012	0.53	0.61	10.93	11.54
	5.1 to Div	340	275	30	0.50	0.013	6.15	0.082	0.012	0.41	0.49	10.18	0.59
	Div to DTW	340	275	60	1.00	0.05	5.9	0.295	0.046	1.56	1.86	11.54	13.4

Pipe friction losses are based on Hydraulics Research graphs of Colebrook White estimates
Source: Sir M. MacDonald and Partners (1989c).

Outlet 5.1 to division

1 × 45° bend	1 × 0.15	=	0.15
1 riser	1 × 1.5	=	1.5
1 inlet valve	1 × 4.5	=	4.5
	Total		**6.15**

Division to deep tube-well (water source)

2 × 22° bends	2 × 0.1	=	0.2
2 straight tees	2 × 0.1	=	0.2
1 inlet valve	1 × 4.5	=	4.5
1 outlet to division	1 × 1	=	1
	Total		**5.9**

Based on the use of 280mm diameter concrete pipe and these minor loss coefficients the hydraulic calculations are summarized in Table A2.3. Estimates of pipe friction losses are based on Colebrook White (Hydraulics Research, 1990) with a roughness height of 0.6mm.

From the design water level it is clear that for the critical path length the water level is very close to the maximum allowable for the pressure tower. It may be worth considering the use of larger diameter pipe for some of the network to reduce the velocity head and friction losses.

Step 5. Selection and design of remaining component structures

Water hammer protection from rapid valve closure

1 Calculate the maximum allowable surge pressure rise.

Given the use of mortar jointed non-reinforced concrete pipe which had been successfully tested to a hydrostatic pressure of 1kg/m², the maximum allowable operating pressure plus surge pressure would be 30% of 10m = 3m.

Given an operating pressure at the open outlet of 0.5m.

Maximum allowable surge = 3 – 0.5 = 2.5m.

2 Select a first guess of minimum distance of outlet from free water surface.

Suggested minimum distance of 200m.

3 Calculate the peak pressure rise occurring at the instantaneous time of closure.

Peak pressure rise given by

$$\frac{V_o \times U}{g}$$

where

$U = 1100$m/s and $V_o = 0.94$m/s

Peak pressure

= 0.94 × 1100/9.81 = 105m at instantaneous closure.

Instantaneous time of closure

= 2L/U = 2 × 200 / 1100 = 0.36s.

4 Calculate pressure rises for slower valve closure

Given valve closure T_1 takes 20s what is the average change in velocity?

Average change in velocity

$= 0.5\ T_c \times V_o\ /\ T_1 = 0.5 \times 0.36 \times 0.94\ /\ 20$

$= 0.008$ m/s

Maximum change in velocity

$= 0.008 \times 3 = 0.025$m/s

Maximum head rise

$= 1100 \times 0.025\ /\ 9.81$

$= 2.8$m

The head rise is in excess of the allowable surge pressure for the pipe to be used. The surge pressure rise can be reduced by any of the following:

- decrease the distance to the free water surface;
- slow down the rate of valve closure (not practical);
- use a pipe material with a higher allowable working pressure.

Assuming the only option is to decrease the spacing between the outlet and the free water surface the analysis is reworked for a spacing of 150m.

In this instance the following results are obtained:

- Instantaneous time of closure equals 0.27s.
- Average change in velocity $= 0.5 \times 0.27 \times 0.94\ /\ 20 = 0.0063$m/s.
- Maximum change in velocity $= 3 \times 0.0063 = 0.019$m/s.
- Peak pressure rise $= 0.019 \times 1100\ /\ 9.81 = 2.13$m.

This is within allowable surge pressure rise and is thus acceptable.

Additional surge risers are positioned such that no outlet is more than 150m from a free water surface, or in this case an open stand or vent. Where open vents cannot be used pressure release valves set at the maximum allowable working pressure should be installed.

In the case of Bhaluka surge risers will be required at the ends of the branches with outlets, 5.1, 7.1 and 6.3. Additional risers need to be placed at positions between outlets 6.1 and 6.2; 1.1 and 1.2; and near outlets 4.3 and 2.2.

Air vent placement

After completion of long sections of the pipe alignment, positions at which air vents are required can be identified. Surge risers may suffice as air vents in many of these positions. Where air vents are specifically required open vents or air release valves could be used.

APPENDIX 3

Worked example for the design of a float valve stand for a semi-closed pipe system

This worked example with relevant explanation is given to help explain the design of float valves and their stands for use in semi-closed pipe systems. The analysis is based on Merriam (1987b), and is necessarily an approximation of actual hydraulic conditions.

The following data is given for the particular section of pipeline to be designed and would be required prior to undertaking a particular design. The design method applies to only one pipe section between two float valve stands and to the design of the lower float valve and stand. Where sections between stands are similar then the analysis could be generalised, however it will usually be necessary to complete the design for each semi-closed pipeline section.

Ground slope (s_g) = 0.01

Design discharge (Q_{des}) = 60 l/s

Non-reinforced concrete pipe to be used, K_s for Colebrook White = 0.6

Safe pressure for concrete pipe (M) = 6m

Minimum energy head (E) = 1.5m

Pipe to have minimum cover of 1m: (D) = 1m.

therefore stand spacing (L) is given by:

$$M = E + D + s_g L$$

or

$$L = M - (E+D)/s_g$$

so that

$$L = (6 - 2.5)/0.01$$

$$L = 300\text{m}$$

Step 1. Pipeline selection

An appropriate pipe size must have a friction loss for the design flow rate which is less than the ground slope. Flow velocity should also be kept below 1.5 m/s, for the design flow rate of 60 l/s.

The following pipe options are available to provide a flow of 60 l/s. This information can be ascertained directly from Colebrook White equation values given in tables prepared by Hydraulics Research (1990).

Pipe diameter (mm)	*Friction loss (m)*	*Velocity (m/s)*
275	*0.0046*	*1.01*
250	*0.0078*	*1.22*
225	*0.0133*	*1.51*

The pipe diameter of 275 mm is chosen for the lower velocity given that the cost of the larger capacity is small in the case of concrete pipes.

Step 2. Calculate head to be dissipated at the float valve

Given the conservation of energy on either side of the float valve the following equation holds true (see Figure A3.1).

$$s_gL + E = sL + h_v + h_t + E$$

where

s_gL = change in elevation
E = minimum energy above ground
sL = pipe friction loss
h_v = velocity head
h_t = head loss by throttling

Thus

$$h_t = s_gL - sL - h_v$$
$$h_t = 3 - 1.38 - 0.05 = 1.57\text{m}$$

Step 3. Select float valve from manufacturer's data

Using the manufacturers' data given in Figure A3.2, the float valve should pass the design flow with a friction loss less than h_t.

With a head difference of less than 8m an unbalanced float valve will be adequate.

The size of unbalanced float valve most suitable is 200mm (0.2m) with a head loss of 0.5m at design flow.

Step 4. Calculate the maximum flow condition for the wide open valve

In this case the entire available head is dissipated through the float valve. The maximum flow is found from the following equation:

$$Q = CA\,(2gh_t)^{0.5}$$

where

C = 0.6 for unbalanced valves
C = 0.95 for balanced valves
h_t = head to be throttled
× 1000 to give l/s

Maximum flow from the graph for h_t = 1.57m is 105 l/s.
This can be verified for agreement by the equation.

From the equation

$$Q = 0.6 \times 0.0314 \times 1000 \times (2 \times 9.81 \times 1.57)^{0.5}$$
$$= 104 \text{ l/s approx } 105 \text{ l/s}$$

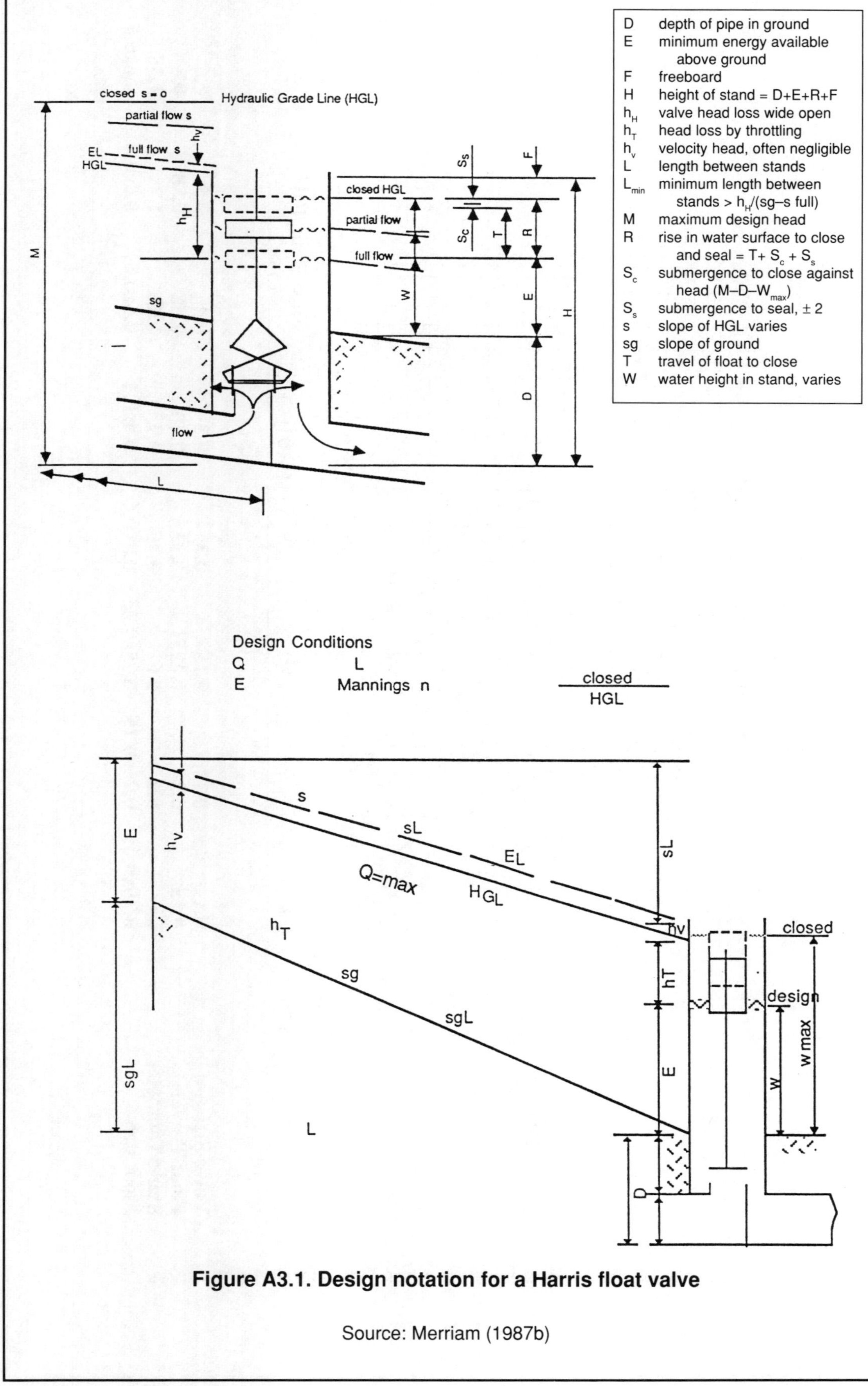

Figure A3.1. Design notation for a Harris float valve

Source: Merriam (1987b)

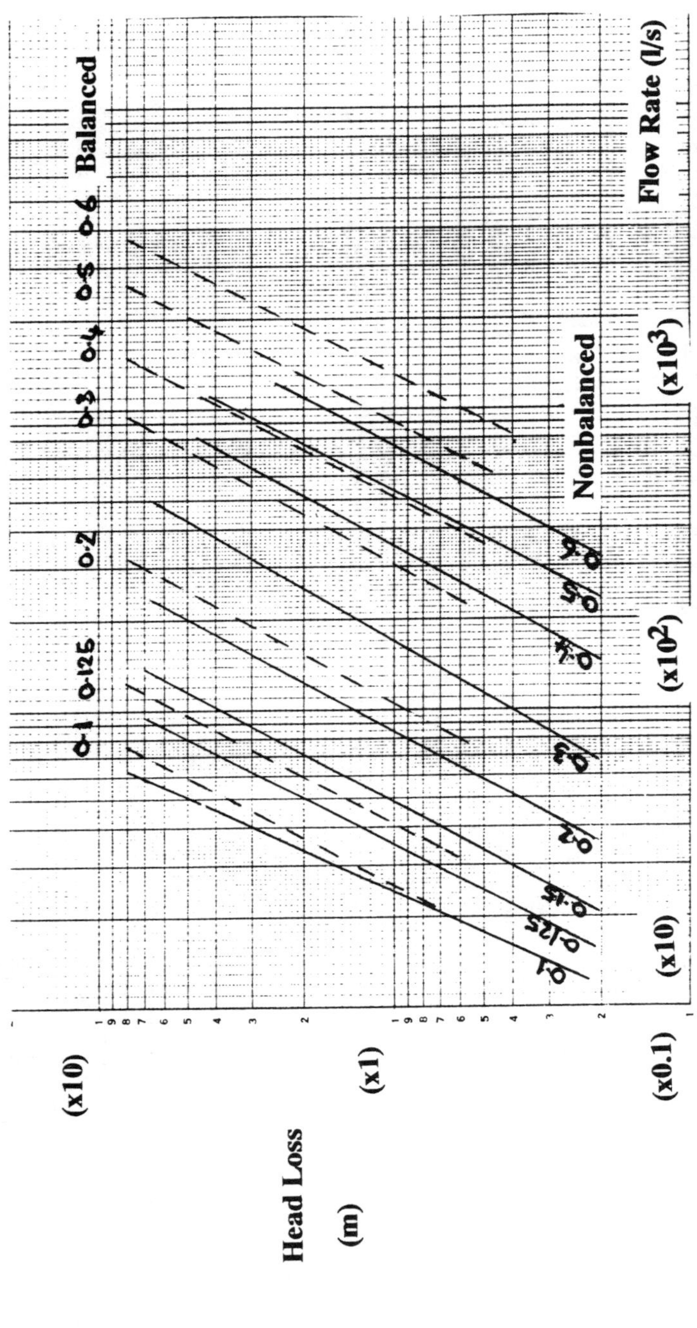

Valve Size (m)	0.1	0.125	0.15	0.2	0.3	0.4	0.5	0.6
Float Travel	0.15	0.2	0.225	0.33	0.48	0.58	0.58	0.8
Stand Diameter	0.6	0.6	0.6	0.75	0.9	1.05	1.2	1.35
Float size	0.25x0.25	0.25x0.25	0.25x0.25	0.4x0.4	0.4x0.4	0.6x0.6	0.6x0.6	0.6x0.6

Figure A3.2. Head loss for wide-open-flow Harris float valves

Source: Merriam (1987b)

Step 5. Establish the float setting for the design flow through the float valve

The valve cross-section is equivalent to the same pipe diameter, and the valve is considered to be wide open when the opening is one quarter of the pipe diameter. The float travel distance from fully open to closed (T) is defined by the manufacturer for each valve and float combination and given in Figure A3.2.

Full float travel distance (T) for a 200 mm unbalanced float valve = 0.33m

The required float position to deliver the design flow is determined by the ratio of Q_{des}:Q_{max} where the design float travel (T_1) is equal to full float travel (T)× the ratio of closure (R_c).

The ratio of valve closure (R_c) will equal 1–(Q_{des}:Q_{max})

therefore

$$R_c = 1 - (60/105) = 0.43$$

Design float travel

$$(T_1) = T \times R_c$$

therefore

$$T_1 = 0.43 \times 0.33 = 0.14\text{m}$$

The level of the float bottom with the valve wide open will be set by the design head (E) less the float travel to achieve design throttling (T_1) and also less the submergence to support the weight of the float and mechanism (S_v).

Float bottom level = E – Float travel at design (T_1) – Submergence (S_v)

S_v = submergence to support weight of valve body and float, usually in the range 0.05–0.15m depending on the size of the valve.

where submergence (S_v) is generally = 0.07m though it may vary for different valve and float combinations.

Therefore

float bottom level = 1.5 – 0.14 – 0.07 = 1.29m

Step 6. Calculate submergence to close, S_c

In order to calculate the maximum water level in the float valve it is necessary to calculate submergence to close at or near no flow. The head of water on either side of the float valve can be described by the following equations for both upstream and downstream conditions.

Upstream water head

$$= M = W + D + sgL$$

While downstream head

$$= D + W_{max}$$

At low or no flow conditions D (depth of cover over the pipe) and W (water level in the stand) are the same in both the top and bottom stands on the pipe section. The head difference across the valve can then be represented by the term sgL (the change in elevation). It is this force which resists the closing of the float valve and must be overcome by an equal and opposing force supplied by the rising water level on the float.

The uplift force opposing valve closure in kg/m^3 = $A \times s_g L \times 1000$

In this case the uplift force = $0.0314 \times 3 \times 1000 = 94.2$ kg/m^3

Because of the lever arm action the rising float is only required to supply an uplift force equivalent to one sixth of the uplift force to achieve closure.

Required float lift at closure

$$= \frac{A \times s_g L \times 1000}{6}$$

For this worked example the required float lift at closure = 94.2/6 = 15.7 kg/m^3

The float lift is provided by the buoyancy of the polystyrene float. Each cubic metre of float will provide a lift of 850 kg force The float lift for each metre height of the float will depend on the cross section of the float and is given by the equation which follows.

Unit float lift = float area × unit height (m) × unit float lift 850 kg/m^3

Therefore

Unit float lift = $0.0625 \times 1 \times 850 = 53.12$ kg lift/m

and

$$S_c = \frac{\text{required uplift force at closure}}{\text{float lift (kg lift/m)}} = 15.7/53.12 = 0.3\text{m}$$

With submergence to close (S_c) known it is then possible to calculate W_{max} using step 7.

It is also possible to calculate the height of the float required from the information already obtained.

float height = $S_c + S_v + S_s$

In this instance required float height assuming a cross-section of 0.0625 m^2 will be:

0.3 + 0.07 + 0.07 = 0.44m.

In order to provide the correct float for each stand, the polystyrene float can be trimmed to the required dimensions and volume.

Step 7. Calculate maximum water level in the float stand

Maximum water level W_{max} will occur at valve closure and sealing. It will equal the minimum design level (E) plus the remaining valve travel distance from design to closure and an allowance for the submergence of the float to close and seal the valve against the pressure difference. This can be represented by;

$$W_{max} = E + (T-T_1) + S_c + S_s$$

where

T_1 = float rise required for design flow
S_c = submergence to close
S_s = submergence to seal and is usually set at 0.07m

Thus

$$Wmax = 1.5 + (0.33-0.14) + 0.3 + 0.07 = 2.06m$$

Step 8. Stand height and maximum upstream pressure

Stand height will equal the maximum water level plus some allowance for freeboard (typically 0.3) and the embedment depth of the pipe. The freeboard can be adjusted by changing the float setting and the minimum desired head E since the value of the rise remains constant.

Stand height
$$= Wmax + Freeboard + D$$
Stand height
$$= 2.06 + 0.3 + 1 = 3.36m$$

The maximum upstream pressure head on the pipe will be;

$$Wmax + s_g L + D$$

where

D = depth of pipe in ground

In this pipe section maximum upstream pressure
$$= 2.06 + 3 + 1 = 6.06m$$

This is considered acceptable though if the pressure was considered excessive then the freeboard could be reduced to bring the maximum upstream pressure to below 6 m.

Thus we see that a structure of about 2 metres height functions to control maximum water pressures of up to 6 metres. This would not be possible using an open stand type of structure. However pipes which can withstand higher maximum pressures also become necessary.

Where the maximum pressure exceeds the safe pipe pressure it can also be reduced by;

- Closer spacing of the stands as long as sufficient head is available at design flow for valve friction losses.
- The minimum desired head E can be reduced by lowering the bottom float setting.

REFERENCES

Adams, W.M. (1992.) *Wasting the rain. Rivers, people and planning in Africa.* Earthscan Publications Ltd. London. UK.

ACI. (1980.) 'Recommendations for Cast-in-Place Nonreinforced Pipe'. Report from ACI Committee 346. American Concrete Institute, ACI 346R-81.

ASAE S261.7. (1989.) 'Design and Installation of Nonreinforced Concrete Irrigation Pipe Systems'. American Society of Agricultural Engineers, American National Standards Institute. ANSI/ASAE S261.7. USA.

ASAE S376.1. (1989.) 'Design, installation and performance of underground, thermoplastic irrigation pipelines'. American Society of Agricultural Engineers, American National Standards Institute. ANSI/ASAE S376.1. USA.

ASTM. 'Specifications for concrete pipe for irrigation and drainage'. C118. American Society of Testing Materials.

Baudequin, D. , Galand, A. and Renault, D. (1990.) 'Les reseaux basse-pression en irrigation de surface'. (Low pressure pipe networks for surface irrigation) CEMAGREF/ S.C.P./ENGREF. France

Bealey, M. (1987.) 'Durability considerations: precast concrete pipe'. Katherine and Bryant Mather International Conference on Concrete Durability. Michigan, USA.

van Bentum, R.J. (1992.) 'Low pressure buried pipe distribution systems for surface irrigation'. Unpublished MPhil thesis. Loughborough University of Technology. UK.

van Bentum, R.J. and Smout, I.K. (1993.) 'Photographs of Buried Pipelines for Surface Irrigation'. Water, Engineering and Development Centre, Loughborough University of Technology. UK.

van Bentum, R.J. and Smout, I.K. (1990.) Interim report on desk study. 'Research into buried pipe distribution systems for irrigation'. Overseas Development Administration (UK) Project R4575: Water, Engineering and Development Centre, Loughborough, UK.

van Bentum, R.J. and Smout, I.K. (1991.) 'Supporting data collected during study tour and research project'. Overseas Development Administration, UK, Project R4575. WEDC. Loughborough, UK.

Booher, L. J. (1974.) *Surface irrigation.* Food and Agricultural Organization of the United Nations (FAO). FAO Agricultural Development Paper No. 95. FAO/University of California, Davis, USA.

Bos, M.G. (1985.) 'Summary of ICID definitions on irrigation efficiency'. January 1985. ICID Bulletin Vol. 34 No. 4.

Bos, M.G. and Ngutren, J. (1990.) *On irrigation efficiencies*. International Institute for Land Reclamation and Improvement. 4th edition. The Netherlands.

Campbell, D.E. (1984.) 'Pipe distribution systems in groundwater development'. Agricultural Administration Vol. 16, 209-227.

Campbell, D.E. (1986.) 'Design and operation of irrigation systems for smallholder agriculture in south Asia'. FAO Investment Centre Technical Paper 3/2. Rome, Italy.

Carruthers, I. and Clark, C. (1981.) *The economics of irrigation*. Liverpool University Press. UK.

Chambers, R. (1988.) *Managing canal irrigation, practical analysis from south Asia*. Cambridge University Press. UK.

Chambers R.E. and McGrathT.J. (1981.) 'Structural design of buried plastic pipe'. Paper 2, Session 1. Proceedings of the International Conference on Underground Plastic Pipe. March 30-April 1, 1981. American Society of Civil Engineers/ Plastic Pipes Institute. New Orleans, USA.

Chandler, J.C. (1987.) 'Case studies, semi-closed pipeline systems. Orange Cove and Solano Irrigation District'. Symposium on Planning, Operation, Rehabilitation and Automation of Irrigation Water Delivery Systems. ASCE Irrigation and Drainage Division Specialty Conference. Oregon, USA.

Chengzhi, Z. (September 1988.) 'Irrigation system with underground plastic hose and the experiments'. Proceedings of International Conference on Irrigation System Evaluation and Water Management. Wuhan, China.

Coles, E. D. (1991.) Comments on low pressure pipeline systems. Personal correspondance regarding consulting experience with low pressure buried pipelines. Unpublished.

Coward, E.W. (editor). (1980.) *Irrigation and agricultural development in Asia. Perspectives from the social sciences*. Cornell University Press. Ithaca, USA.

Cunningham, J.F. (1986.) 'Groundwater development technology, an Indian case study The Second Uttar Pradesh Public Tube-wells Project'. World Bank Irrigation/Drainage Seminar.

Deacon, N.H.G. (1984.) 'Seepage and durability of irrigation canal linings, A review of published data'. Tech Note OD/TN5. Hydraulics Research, Wallingford, UK.

DST. (1990.) Irrigation techniques with low pressure pipe distribution. Department of Scientific Technology and Education of Ministry of Water Resources. China.

Doorenbos, J. et al. (1979.) 'Yield response to water'. Irrigation and Drainage Paper No. 33. Food and Agriculture Organization of the United Nations. Rome, Italy.

Dupriez, H. and DeLeener, P. (1992.) *Ways of water; runoff, irrigation and drainage*. Land and Life Series.

REFERENCES

EWUP. (1983.) Design report El-Hammami pipeline, Technical Report No. 21. Fort Collins Staff Team. Egypt Water Use and Management Project. Egypt.

Eisenhauer, D.E. (1990.) Gated pipelines for improvement of surface irrigation efficiency. Siphon No. 5. pp. 10-14. Newsletter from the Centre for Irrigation Engineering, K.U. Leuven. The Netherlands.

Elsevier. (1989.) *Agricultural compendium for rural development in the tropics and sub-tropics.* (Third edition.) Elsevier Scientific Publishing Company. The Netherlands.

Everite Eternit Tac Pressure Pipes Ltd. Everite. (1986.) Manufacturer's literature.

Field, W.P. (1990.) *World irrigation. Irrigation and Drainage Systems* 4. pp. 91-107. Kluwer Academic Publishers. The Netherlands.

Galand, A. (1989.) 'Automation of surface irrigation using the system trans-irrigation: experience and recent developments'. (Colloque sur les methodes d'irrigation ameliorees.) Southern France.

Georgi, F. (1989.) 'Manual on buried pipe irrigation systems'. Tangail Agricultural Development Project. BRDB/GTZ. Bangladesh.

Gisselquist, D. (1986a.) 'Low-cost concrete pipe irrigation system for shallow tube-wells, deep tube-wells and low-lift pumps'. Draft paper. Bangladesh.

Gisselquist, D. et al. 1986b. 'Report on water management study tour to India'. Tangail Agricultural Development Project. BRDB/GTZ. Bangladesh.

Gisselquist, D. (1989.) 'Demonstrating command area development'. Tangail Agricultural Development Project. BRDB/GTZ. Bangladesh.

Goldsmith, H. and Makin, I.W. (1989.) 'Canal Lining: from the laboratory to the field and back again'. Paper 4.3. Proceedings of the International Conference "Irrigation: Theory and Practice." Southampton, UK.

Goussard, J. (1987.) 'Neyrtec automation equipment for irrigation canals. Symposium on Planning, Operation, Rehabilitation and Automation of Irrigation Water Delivery Systems'. ASCE Irrigation and Drainage Division Specialty Conference. Oregon, USA.

GDC. (1987.) Study of groundwater development strategies for irrigation in the Terai (Nepal). Groundwater Development Consultants Int. Ltd. Vol. 4, Engineering and Volume 1, Main Report. Cambridge, UK.

Gumbel, J.E., O'Reilly, M.P., Lake, L.M., and Carder, D.R. (1982.) 'The development of a new design method for buried flexible pipes'. Europipe 1982 Conference, Basel, Switzerland, Paper 8. pp.87-98

Harris. (1990.) 'Float valve literature and costs'. Harris Float Valves Inc. Missoula, Montana, USA.

HPPA. (1989.) 'Ultra-Rib, design and installation guide'. Technical Report. High Performance Pipe Association. Aylesford, UK.

HHP/HTS. (1988.) Sukhothai Groundwater Development Project. Final report. Howard Humphreys and Partners in association with Hunting Technical Services. Royal Irrigation Department, Kingdom of Thailand.

Hydraulics Research. (1990a.) *Tables for the hydraulic design of pipes and sewers*. Fifth edition. Wallingford, UK.

Hydraulics Research. (1990b.) *Charts for the hydraulic design of channels and pipes*. Sixth edition. Hydraulics Research Station, Wallingford, UK.

Hydraulics Research. (1990c.) 'Proposal for research into the performance of low pressure pipe irrigation systems in Egypt'. Research Proposal OD/RP 74. Wallingford, UK.

IDP WRC. (1990.) Experiences in design and construction of two buried pipe schemes in Faridpur and Kurigram. IDP Water Resources Cell. Rural Employment Sector Programme. Bangladesh.

Indian Standards 783. (1985.) 'Standard for the construction and testing of non-pressure concrete buried pipelines using rigid and flexible joints'. Indian Standards Association. New Delhi.

IDTP. (1989.) Indo-Dutch Tube-well Project Monitoring and Appraisal Cell. Indo-Dutch Uttar Pradesh Tube-well Project. Draft inception report. Lucknow, Uttar Pradesh, India.

Jain. (1989.) 'Ribloc PVC pipes for low pressure irrigation'. Short report. Jain Group of Industries. Jalgaon, Maharashtra, India.

Jain. (1990.) 'Technical specification of lightweight ribbed PVC pipe with smooth inside and helical "T" outside. Draft publication. Jain Group of Industries, India. Jalgaon, India.

Jain Ribloc. (1991.) 'Installation of Ribloc pipe, users manual and Ribloc design guidelines'. Jain Plastics and Chemicals Pvt. Ltd. Ribloc Division. Jain Group of Industries. Jalgaon, India.

James, L.G. (1988.) *Principles of farm irrigation design*. Washington State University. Wiley and Sons, USA.

Järvenkylä, J.J. (1988.) 'Thin-walled thermoplastic pipes'. Paper 30. International Conf. Plastics Pipes VII, 19-22 September 1988. University of Bath, UK.

Keller, J. (1990.) 'Modern irrigation in developing countries'. Fourteenth Congress of the International Commission on Irrigation and Drainage. Special Session R. 9, Rio de Janeiro, Brazil.

Kolavalli, S. and Shah, N. (1989.) 'Management of public tube-wells in Uttar Pradesh'. Draft report. Centre for Management in Agriculture. Indian Institute of Management, Aurangabad, India.

Koluvek, P. K. (1970.) Report to the government of India on design criteria, construction guide, and material standards for irrigation pipelines. USAID Mission to India.

Kruse, E.G., A.S. Humphreys, and E.J. Pope. (1980.) 'Farm water distribution systems'. In: Jensen, M.E. (ed), *Design and Operation of Farm Irrigation Systems*. Monograph No. 3. American Society of Agricultural Engineers, St Joseph, Michigan.

Labye, Y., Olson, M.A., Galand, A. and Tsiourtis, N. (1988.) 'Design and optimization of irrigation distribution networks'. FAO Irrigation and Drainage Paper 44. Food and Agricultural Organization of the United Nations, Rome, Italy.

Manuellan G. (1988.) 'Developments in water conveyance and water control in France. Technological and Institutional Innovation in Irrigation'. Proceedings of a Workshop held at the World Bank. World Bank Technical Paper No. 94. Part II. Washington, USA.

Merriam, J.L. (1985.) 'Demand irrigation schedule concrete pipeline pilot project'. Final report. Mahaweli Development Board, Area H. Sri Lanka.

Merriam, J.L. (1987a.) 'Case study Open pipeline system. Coachella Valley Water District'. Symposium on Planning, Operation, Rehabilitation and Automation of Irrigation Water Delivery Systems. ASCE Irrigation and Drainage Division Specialty Conference. Oregon, USA.

Merriam, J.L. (1987b.) 'Design of semi-closed pipeline systems'. Symposium on Planning, Operation, Rehabilitation and Automation of Irrigation Water Delivery Systems. ASCE Irrigation and Drainage Division Speciality Conference. Oregon, USA.

Merriam, J.L. (1987c.) 'Pipelines for flexible deliveries'. Symposium on Planning, Operation, Rehabilitation and Automation of Irrigation Water Delivery Systems. ASCE Irrigation and Drainage Division Speciality Conference. Oregon, USA.

Merriam, J.L. (1987d.) 'Reservoirs help on-farm operation and automation'. Symposium on Planning, Operation, Rehabilitation and Automation of Irrigation Water Delivery Systems. ASCE Irrigation and Drainage Division Speciality Conference. Oregon, USA.

Merriam, J.L. (1990.) 'Gadigaltar tank irrigation pilot project'. Irrigation Department, Khargone, Madhya Pradesh, India.

Michael, A. M. (1978.) 'Irrigation theory and practice'. Vikas, New Delhi, India.

Mock, J.F. and Bolton, P. (1991.) 'Environmental effects of irrigation, drainage and flood control projects. Checklist for environmental impact indication'. Hydraulics Research Report OD/TN 50. Hydraulics Research Overseas Development Unit.

MMI. (1992.) Deep Tubewell II Project. Final Report. Main Report and Supplement 2.4/2 Bangladesh Agricultural Development Corporation, Bangladesh.

MMI. (1990a.) S.I. Brod: 'Tour report, 16 December 1989 - 14 June 1990'. IDA Deep Tubewell II Project. Mott MacDonald International Ltd. Bangladesh Agricultural Development Corporation. Bangladesh.

MMI. (1990b.) S.J. Ray: 'Tour report and Annex, 9 January 1989 - 31 July 1990'. IDA Deep Tube-well II Project. Mott MacDonald International Ltd. Bangladesh Agricultural Development Corporation. Bangladesh.

MMP. (1989a.) 'Buried pipe irrigation distribution systems implementation note'. Draft Working Paper No. 34. IDA Deep Tube-well II Project. Sir M. MacDonald and Partners Ltd. Bangladesh Agricultural Development Corporation. Bangladesh.

MMP. (1989b.) 'Groundwater irrigation manual. Assistance in the establishment of design criteria and manuals for irrigation projects in Nepal'. Sir M. MacDonald and Partners Ltd. Cambridge, UK.

MMP. (1989c.) 'The case for buried pipe distribution systems'. IDA Deep Tube-well II Project. Sir M. MacDonald and Partners Ltd. Bangladesh Agricultural Development Corporation. Bangladesh.

MMP. (1989d.) S.A. Waters: 'Tour Report, 23 January - 27 June 1989'. IDA Deep Tube-well II Project. Sir M. MacDonald and Partners Ltd. Bangladesh Agricultural Development Corporation. Bangladesh.

MMP. (1989e.) O.C.C. Taylor: 'Tour Report, 23 January - 27 June 1989'. IDA Deep Tube-well II Project. Sir M. MacDonald and Partners Ltd. Bangladesh Agricultural Development Corporation. Bangladesh.

Mridha, M.A.K. (1993.) 'The performance of concrete buried pipe distribution systems for irrigation under farmer management in Tangail, Bangladesh'. Unpublished M.Phil. thesis. Loughborough University of Technology. UK.

Ostrom, E. (1992.) 'Crafting institutions for self-governing irrigation systems'. Institute for Contemporary Studies. San Francisco, USA.

Pant, N. (1989.) 'Water distribution and management in new design public tube-wells in eastern Uttar Pradesh'. MAC Indo-Dutch Tube-well Project and Centre for Development Studies. Lucknow, India.

Pillsbury, A.F. (no date.) *Concrete pipe for irrigation*. University of California Publications. USA.

Pimley, L. and Fischer, G. (1990.) 'Design, installation and maintenance of pipe systems'. Session 8-1, Water Systems Operation and Maintenance Workshop. US Department of the Interior, Bureau of Reclamation. Denver, Colorado, USA.

Plusje, J. (1981.) 'Short report of experimental underground low pressure pipe system for irrigation water distribution'. Bogra District. Bangladesh.

Plusquellec, H. et al. (1988.) 'Technological and Institutional Innovation in Irrigation'. Proceedings of a Workshop held at the World Bank. World Bank Technical Paper No. 94. Washington, USA.

Prokashali Sangsad Consulting Engineers. (1987.) Design report prepared for Barind Integrated Agricultural Development Project. Dhaka, Bangladesh.

Rashid, M H., Mridha, M.A.K., van Bentum, R.J. and Smout I.K. (1990a.) 'Research into buried pipe distribution systems for irrigation, interim fieldwork report: 1989-90 Irrigation Season'. Overseas Development Administration, UK. Project R4575. Bangladesh Agricultural Research Institute/Water, Engineering and Development Centre. Loughborough, UK.

Rashid, M.H., Hye, A.K.M. and Das, R.K. (1990b.) 'Evaluation of traditional (farmer-managed) deep tube-well irrigation systems (1988-89 irrigation season)'. IDA-DTW II Project. Agricultural Engineering Division, BARI. Joydepur, Gazipur, Bangladesh.

Rashid, M.H., Mridha, M.A.K., van Bentum, R.J. and Smout I.K. (1992.) Research into buried pipe distribution systems for irrigation, final report, 1989 to 1991: Studies on eight deep tube-wells in Tangail district, Bangladesh. Overseas Development Administration UK, Project R4575. Bangladesh Agricultural Research Institute / Water, Engineering and Development Centre. Loughborough, UK.

Renault, D. (1988.) *Modernization of furrow irrigation in the south-east of France, automation at field level and its implications*. Irrigation and Drainage Systems 2: 229-240. Kluwer Academic Publishers. Dordrecht, Netherlands.

Republic of Indonesia. (1990.) 'Groundwater development project in east Java. The utilization of buried pipeline for irrigation distribution system'. Directorate of Irrigation II. Ministry of Public Works. East Java, Indonesia.

Riblok, Australia, Ltd. (1990.) Product literature, covering material use in agriculture, for concrete formwork, stormwater drainage, sewer relining and concrete pipe formwork.

Rotacurve. (no date.) 'An alternative bend for tongue and groove pipes'. California, USA.

Rouke, D. (1984.) 'Pipelines (design, construction and operation)'.

Rushton, K.R. et al. (1992.) 'Interim results of numerical model study of losses from lined canals'. SERC Grant No. GR/F/05565. University of Birmingham. UK.

Sargardoy et al. (1982.) 'Organization, operation and maintenance of irrigation schemes'. FAO Irrigation and Drainage Paper No. 40. Food and Agriculture Organization of the United Nations. Rome, Italy.

Seckler, D. (1986.) Institutionalism and agricultural development in India. Journal of Economic Issues. Vol. XX, No. 4.

Smout, I. (1990.) 'Farmer participation in planning, implementation and operation of small-scale irrigation projects'. ODI-IIMI Irrigation Management Network. Paper 90/2b.

Stern, P. (1979.) *Small-scale irrigation: A manual of low-cost water technology*. Intermediate Technology Publications Ltd. and International Irrigation Information Centre. Israel.

Tiffen, M. (1989.) 'Guidelines for the incorporation of health safeguards into irrigation projects through intersectoral cooperation'. Joint WHO/FAO/UNEP panel of experts on environmental management for vector control (PEEM). Guidelines series 1.

Titow, W.V. (1985.) *PVC technology*. Fourth edition. Elsevier Applied Science Publishers. London.

Trout, T.J. and Kemper, W.D. (1980.) 'Watercourse improvement manual'. Water management technical report no. 58. Consortium for International Development. Water Management Research Project. Colorado State University. USA.

Twort, A.C., Law F.M. and Crowley, F.W. (1985.) *Water supply*. Third edition. Edward Arnold. London.

US SCS. (1967.) *National Engineering Handbook* Section 2, Part 1. Engineering Conservation Practices. United States Soil Conservation Service.

Univ. Cal. (1977.) Leaflet 2908, 'Low-head irrigation pipe: concrete, asbestos-cement, and plastic'. University of California Division of Agricultural Sciences. Davis, USA.

...or designing and evaluating surface irrigation systems'.
...n of the United Nations. Irrigation and Drainage Paper

... 'The long-term behaviour of buried uPVC sewer pipe'.
...s Pipes VII, 19-22 September 1988, University of Bath, UK.

...ried pipe water distribution system'. An interim report.
... Project. Water and Power Development Authority of
...ID and University of Idaho, USA.

...rrigation appliance catalogue A. Red Top water control
...ter, California, USA.

...ndustry for the structural design of underground non-
...Research Centre. Swindon, UK.

...ign, installation, and operation of uPVC pressure pipe
... Swindon. UK. 1989.

WRC. (1991.) 'Water Industry Specification, Specification for solid wall concentric external rib-reinforced uPVC sewer pipe'. Water Research Centre. (IGN No. 4-31-05.

Watts, P. and Smith, P. (1985.) 'Design of gated pipe irrigation systems'. Fifth Afro-Asian Regional Conference on Irrigation and Drainage. Planning and management of water for agriculture in the tropics. Townsville, Australia.

World Bank. (1983.) Second Uttar Pradesh Public Tube-wells Project. Staff appraisal report. India.
Young, O.C. and O'Reilly, M.P. (1983.) *A guide to design loadings for buried rigid pipes*. Transport and Road Research Laboratory. Department of Transport. HMSO. London, UK.

Young O.C., Brennan G. and O'Reilly M.P. (1986.) *Simplified tables of external loads on buried pipelines*. HMSO, London, UK.